STUDY AND REVISION GUIDE

T0173281

Working for over
25 YEARS
WITH
Cambridge Assessment International Education

Cambridge IGCSE™

....

Physics

Third Edition

....

Mike Folland
Catherine Jones

HODDER
EDUCATION
AN HACHETTE UK COMPANY

This text has not been through the Cambridge International endorsement process. Any references or materials related to answers, grades, papers or examinations are based on the opinion of the author. The Cambridge International syllabus or curriculum framework, associated assessment guidance material and specimen papers should always be referred to for definitive guidance.

Every effort has been made to trace all copyright holders, but if any have been inadvertently overlooked, the Publishers will be pleased to make the necessary arrangements at the first opportunity.

Although every effort has been made to ensure that website addresses are correct at time of going to press, Hodder Education cannot be held responsible for the content of any website mentioned in this book. It is sometimes possible to find a relocated web page by typing in the address of the home page for a website in the URL window of your browser.

Hachette UK's policy is to use papers that are natural, renewable and recyclable products and made from wood grown in well-managed forests and other controlled sources. The logging and manufacturing processes are expected to conform to the environmental regulations of the country of origin.

Orders: please contact Hachette UK Distribution, Hely Hutchinson Centre, Milton Road, Didcot, Oxfordshire, OX11 7HH. Telephone: +44 (0)1235 827827. Email education@hachette.co.uk Lines are open from 9 a.m. to 5 p.m., Monday to Friday. You can also order through our website: www.hoddereducation.com

ISBN: 978 1 3983 6137 9

First published in 2005
Second edition published in 2016
This edition published in 2022 by
Hodder Education,
An Hachette UK Company
Carmelite House
50 Victoria Embankment
London EC4Y 0DZ

www.hoddereducation.co.uk

Impression number   10 9 8 7 6 5 4 3 2 1

Year   2026 2025 2024 2023 2022

Cover photo © Zffoto / stock.adobe.com

Typeset in India

Printed in Spain

A catalogue record for this title is available from the British Library.

Contents

Answers to exam-style questions are available at:
www.hoddereducation.co.uk/cambridgeextras

Introduction

Welcome to the Cambridge IGCSE™ Physics Study and Revision Guide. This book has been written to help you revise everything you need to know and understand for your Physics exam. Following the Physics syllabus, it covers all the key Core and Extended content and provides sample questions and answers, as well as practice questions, to help you learn how to answer questions and to check your understanding.

How to use this book

Key objectives

The key skills and knowledge covered in the chapter. You can also use this as a checklist to track your progress.

Revision activities

Examples of strategies to help you revise effectively.

Skills

Key practical skills coverage will help you to consolidate your understanding of practical work you have undertaken in your lessons, and to describe and evaluate these skills effectively.

Key mathematical skills are covered to help you to demonstrate these skills correctly.

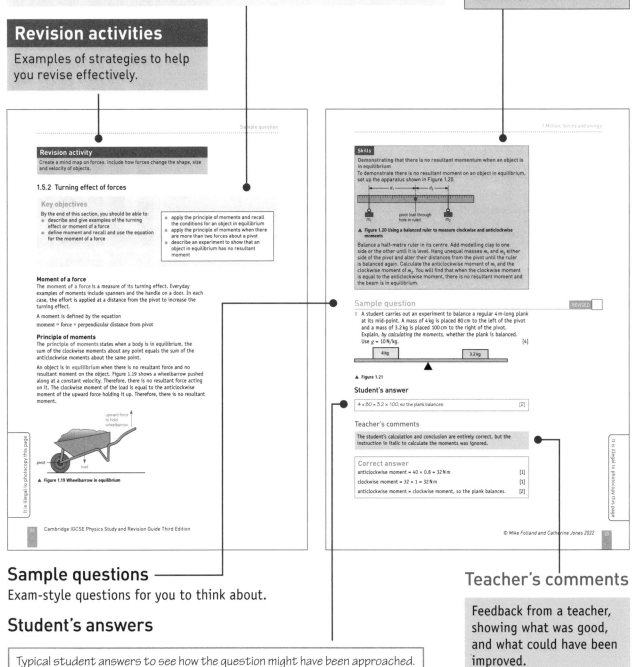

Sample questions

Exam-style questions for you to think about.

Student's answers

Typical student answers to see how the question might have been approached.

Teacher's comments

Feedback from a teacher, showing what was good, and what could have been improved.

Correct answers

Model student answers, based on the teacher's comments on the typical student answers.

Exam-style questions

Practice questions, set out as you would see them in the exam paper, for you to answer so that you can check what you have learned.

Extended syllabus

Content for the Extended syllabus (Supplement material) is shaded yellow.

Answers

Worked answers to the Exam-style questions can be found at:
www.hoddereducation.co.uk/cambridgeextras

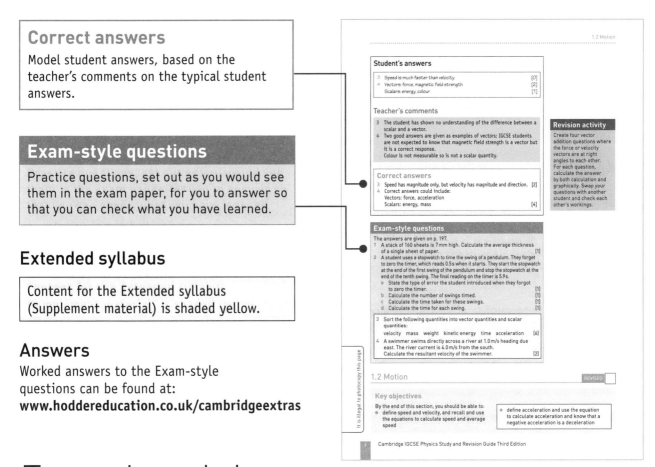

Exam breakdown

You will take three examinations at the end of your studies. If you have studied the Core syllabus content you will take Paper 1 and Paper 3, and either Paper 5 or Paper 6. If you have studied the Extended syllabus content (Core and Supplement) you will take Paper 2 and Paper 4, and either Paper 5 or Paper 6.

Paper 1: Multiple choice (Core)	Paper 3: Theory (Core)
45 minutes	1 hour 15 minutes
40 marks	80 marks
40 four-option, multiple-choice questions based on the Core subject content	Short-answer and structured questions based on the Core subject content
30% of your grade	50% of your grade

Paper 2: Multiple choice (Extended)	Paper 4: Theory (Extended)
45 minutes	1 hour 15 minutes
40 marks	80 marks
40 four-option, multiple-choice questions, based on the Core and Supplement subject content	Short-answer and structured questions, based on the Core and Supplement subject content
30% of your grade	50% of your grade

Paper 5: Practical test	Paper 6: Alternative to practical
1 hour 15 minutes	1 hour
40 marks	40 marks
Questions will be based on the experimental skills in Section 4	Questions will be based on the experimental skills in Section 4
20%	20%

How to prepare for your examinations

Here are a few summary points to guide you:

- Use this book – it has been written to help students achieve high grades.
- Learn all the work – low grades are nearly always attributable to inadequate preparation. If you can recall the work and show understanding of it, you will succeed. Do not leave things to chance.
- Practise skills such as calculations, equation writing, labelling diagrams and the interpretation of graphs.
- Use past papers to reinforce revision, to become familiar with the types of question and to gain confidence.
- Answer each question as instructed on the paper – be guided by the key words used in the question (describe, explain, state etc.). Do not accept a question as an invitation to write what you know about the topic.

Examination terms explained

The examination syllabus gives a full list of the command terms used by in the exam and how you are expected to respond. This is summarised below.

Command word	Explanation
Calculate	Work out from given facts, figures or information
Compare	Identify/comment on similarities and/or differences
Define	Give the precise meaning
Describe	State the points of a topic / give the characteristics and main features
Determine	Establish an answer using the information available
Evaluate	Judge or calculate the quality, importance, amount or value of something
Explain	Set out purposes or reasons / make the relationships between things evident / state why and/or how, and support with relevant evidence
Give	Produce an answer from a given source or use recall/memory
Identify	Name/select/recognise
Outline	Set out the main points briefly
Predict	Suggest what may happen, based on available information
Sketch	Make a simple freehand drawing, showing the key features, and taking care over proportions
State	Express in clear terms
Suggest	Apply knowledge and understanding to situations where there is a range of valid responses, in order to make proposals / put forward considerations

Cambridge IGCSE Physics Study and Revision Guide Third Edition

1 Motion, forces and energy

Key terms

Term	Definition
Acceleration of free fall, g	For an object near to the surface of the Earth this is approximately constant and is approximately $9.8\,m/s^2$
Accuracy	An accurate measurement is one that is close to its true value
Air resistance	Frictional force opposing the motion of a body moving in air
Centre of gravity	The point at which all the mass of an object's weight can be considered to be concentrated
Density	The mass per unit volume
Energy	Energy may be stored as kinetic, gravitational potential, chemical, elastic (strain), nuclear, electrostatic or internal (thermal)
Equilibrium	When there is no resultant force and no resultant moment on an object
Extension	Change in length of a body being stretched
Friction	Force which opposes one surface moving, or trying to move, over another surface
Gravitational field strength	The force per unit mass
Mass	A measure of the quantity of matter in an object at rest relative to an observer
Moment of a force	Moment = force × perpendicular distance from the pivot
Non-renewable	Cannot be replaced when used up
Power	The work done per unit time and the energy transferred per unit time
Pressure	The force per unit area
Principle of conservation of energy	Energy cannot be created or destroyed; it is always conserved
Principle of moments	States when a body is in equilibrium; the sum of the clockwise moments about any point equals the sum of the anticlockwise moments about the same point
Random error	Error introduced by the person taking the measurement
Renewable	Can be replaced; cannot be used up
Speed	The distance travelled per unit time
Systematic error	Error introduced by the measuring device
Velocity	Speed in a given direction
Weight	A gravitational force on an object that has mass
Work	A measure of the amount of energy transferred. Work done = force × distance moved in the direction of the force. SI unit is the joule (J)
Acceleration	Change of velocity per unit time
Deceleration	A negative acceleration; velocity decreases as time increases
Efficiency	(useful energy output/total energy input) × 100%
	(useful power output/total power input) × 100%
Impulse	Force × time for which force acts
Limit of proportionality	The point at which the load–extension graph becomes non-linear
Momentum	Mass × velocity
Principle of conservation of momentum	When two or more bodies interact, the total momentum of the bodies remains constant provided no external forces act

Term	Definition
Resultant force	The rate of change in momentum per unit time
Resultant vector	A single vector that has the same effect as the two vectors combined
Scalar	A quantity with magnitude only
Spring constant	Force per unit extension
Terminal velocity	Constant velocity reached when the air resistance upwards equals the downward weight of the falling body
Vector	A quantity which has both magnitude and direction

1.1 Physical quantities and measurement techniques REVISED

Key objectives

By the end of this section, you should be able to:
- describe how to measure length, volume and time using appropriate measuring instruments
- determine an average value for a small distance or short time by measuring multiples
- understand the difference between scalar and vector quantities and give examples of each
- determine the resultant of two vectors of force or velocity at right angles to each other either by calculation or graphically

Each time you measure a quantity you are trying to find its true value. How close you get to the true value is described as the **accuracy** of the measurement.

Length

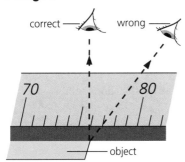

▲ Figure 1.1 The correct way to measure with a ruler

Length is the distance from one end of an object to the other. It is measured using a ruler. To measure length accurately your eye must be perpendicular to the mark on the ruler you are trying to read. This avoids parallax (see Figure 1.1).

Most rulers have millimetre markings. They give values to the nearest mm. For example, if you have to measure a small distance of 4 mm you only know the value to 4±1 mm. To improve this measurement, you measure multiple distances and find an average distance.

Volume

Volume is the amount of space occupied. Figure 1.2 shows how to measure volume using a measuring cylinder. You measure the volume of a liquid by looking at the level of the bottom of the meniscus (see Figure 1.2). (For mercury, you should look at the level of the top of the meniscus.)

Measuring cylinders often measure in millilitres. Remember 1 ml = 1 cm^3.

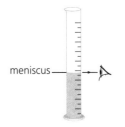

▲ Figure 1.2 The correct way to measure a volume of liquid

Skills

Converting cm³ to m³

The SI unit of length is the metre. It is easy to convert lengths from cm into metres.

$$1\,cm = \frac{1}{100}\,m = 0.01\,m = 1 \times 10^{-2}\,m$$

The SI unit of volume is m³. When you convert the units of volume, you have to divide by 100 for each dimension.

$$1\,cm^3 = \frac{1}{100 \times 100 \times 100}\,m^3 = 0.000\,001\,m^3$$
$$= 1 \times 10^{-6}\,m^3$$

Time

You need to be able to use analogue and digital stopwatches or clocks to measure time intervals. To improve the accuracy of the measurement of a short, repeated time interval, you can measure multiple times. For example, measuring the period of the pendulum in Figure 1.3. The period is the time taken for the pendulum to move from A to B and then back to A. You would measure the time for 10 such swings and then divide the time by 10.

Errors in measurements

In any measurement there may be a measurement error. This is why results are not always the same. The error might be **random** (a **random error**) and cause an **anomaly** when you repeat the result. For example, an error introduced by your reaction time as you start and stop a stopwatch. The error may be a **systematic error**. For example, a newton meter might have a reading even when there is no force applied. This type of error is a zero error. In this case the same error is introduced to all the readings.

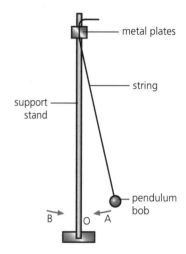

▲ **Figure 1.3 A pendulum**

Scalars and vectors

Quantities can be divided into **scalar** or **vector**:

Scalars:

- only have magnitude (size)
- are added by normal addition

Examples of scalars you should know are distance, time, mass, speed, energy and temperature.

Vectors:

- have direction and magnitude (size)
- are represented by an arrow – the length of the arrow shows the magnitude of the vector and the direction shows the direction it acts
- are added by taking into account their direction

Examples of vectors you should know are force, weight, velocity, acceleration, momentum, electric field strength and gravitational field strength. Always state the direction of a vector, for example, the velocity is 10 m/s northwards.

Finding the resultant of two vectors

When you add vectors, you find the **resultant vector**. This is a single vector that has the same effect as the vectors you have combined. For example, if a force of 200 N pulls a boat to the east and a force of 800 N pulls it to the west, the resultant force is 600 N to the west. You may be asked to find the resultant of two vectors perpendicular to each other. Figure 1.4 shows a 10 N and a 5 N force acting at right angles to each other and their resultant vector F. The forces have been drawn using the parallelogram method and the triangle method.

a
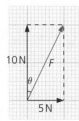

b
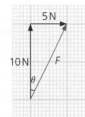

Draw the vectors so they start at the same point. Draw in the sides of the parallelogram. The resultant is the diagonal.

Draw the vectors nose-to-tail. The resultant vector is the line drawn from the start to the finish.

▲ **Figure 1.4 Finding the resultant of two forces acting at right angles to each other**

As you can see from Figure 1.4, the resultant is the same whether you use the parallelogram method or the triangle method. Use whichever one you find easiest. There are two ways to find the value of the resultant vector:

1 By calculation: As it is a right-angled triangle, you can use Pythagoras' theorem to determine the magnitude of the vector.

$$F = \sqrt{10^2 + 5^2} = 11 \text{ N to 2 s.f.}$$

You can use trigonometry to find the angle θ.

$$\tan \theta = \frac{5}{10}, \theta = 27° \text{ to 2 s.f.}$$

Therefore, the resultant is an 11 N force acting at an angle of 27° to the 10 N force.

2 Graphically: By drawing the vectors to scale, you can then use a ruler to measure the length of the resultant vector. Do not forget to convert back using your scale and always write your scale down, e.g. 1 cm : 10 N. You can measure the angle using a protractor. Check this gives the same answer for the resultant force as by calculation using Figure 1.4.

Sample questions

REVISED

1 A student wishes to time how long it takes a ball to fall 1.5 m. Describe how to obtain reliable results for the measurements of time and height. [4]

Student's answer

Start the stopwatch as soon as the ball is released and stop when the ball hits the floor. Repeat three times. If there is an anomalous result, leave it out when you calculate the average time. Measure the height using a ruler and make a mark so that the ball is dropped from the same height each time. [3]

Teacher's comments

The question is answered well in terms of measuring the time. The student realises that they have to repeat because of the random errors when timing. They need to be more detailed in describing the distance measurement. The height is 1.5 m so a tape measure would be needed. It is not accurate to use two smaller rulers. Though they do include the idea of a mark so that the height is the same each time.

Cambridge IGCSE Physics Study and Revision Guide Third Edition

Correct answer

Use a tape measure to make a mark at the correct height of 1.5 m. Release the ball from this height each time. Start the stopwatch when the ball is released and stop it when the ball hits the floor. Repeat the experiment and discard any anomalous results before calculating the average time.   [4]

2 An aircraft flies at 900 km/h heading due south. There is a crosswind of 150 km/h from the west. Graphically, find the aircraft's resultant velocity.   [4]

Student's answer

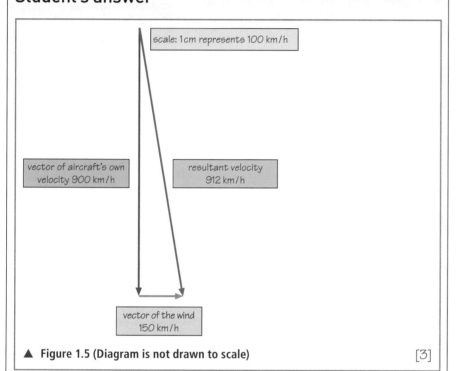

scale: 1 cm represents 100 km/h

vector of aircraft's own velocity 900 km/h

resultant velocity 912 km/h

vector of the wind 150 km/h

▲ **Figure 1.5 (Diagram is not drawn to scale)**   [3]

Teacher's comments

On the whole, the question is extremely well answered and the graphical work is accurate; stating the scale shows excellent work. The student assumed the top of the page was north. However, the student has omitted the direction part of the resultant velocity, stating only the magnitude.

Correct answer

Figure 1.5 should have an arrow pointing up the page labelled north. The answers shown in Figure 1.5 are correct except that the resultant velocity label should be resultant velocity 912 km/h at 9° east of due south.   [4]

3 Speed and velocity are related quantities. Explain why speed is a scalar quantity and velocity is a vector.   [2]
4 Name two more scalar quantities and two more vectors.   [4]

Student's answers

> 3 Speed is much faster than velocity. [0]
> 4 Vectors: force, magnetic field strength [2]
>   Scalars: energy, colour [1]

Teacher's comments

> 3 The student has shown no understanding of the difference between a scalar and a vector.
> 4 Two good answers are given as examples of vectors; IGCSE students are not expected to know that magnetic field strength is a vector but it is a correct response.
>   Colour is not measurable so is not a scalar quantity.

Correct answers

> 3 Speed has magnitude only, but velocity has magnitude and direction. [2]
> 4 Correct answers could include:
>   Vectors: force, acceleration
>   Scalars: energy, mass [4]

Revision activity

Create four vector addition questions where the force or velocity vectors are at right angles to each other. For each question, calculate the answer by both calculation and graphically. Swap your questions with another student and check each other's workings.

Exam-style questions

Answers available at: www.hoddereducation.co.uk/cambridgeextras

1 A stack of 160 sheets is 7 mm high. Calculate the average thickness of a single sheet of paper. [1]

2 A student uses a stopwatch to time the swing of a pendulum. They forget to zero the timer, which reads 0.5 s when it starts. They start the stopwatch at the end of the first swing of the pendulum and stop the stopwatch at the end of the tenth swing. The final reading on the timer is 5.9 s.

 a State the type of error the student introduced when they forgot to zero the timer. [1]
 b Calculate the number of swings timed. [1]
 c Calculate the time taken for these swings. [1]
 d Calculate the time for each swing. [1]

3 Sort the following quantities into vector quantities and scalar quantities:

 velocity   mass   weight   kinetic energy   time   acceleration [6]

4 A swimmer swims directly across a river at 1.0 m/s heading due east. The river current is 4.0 m/s from the south.
 Calculate the resultant velocity of the swimmer. [2]

1.2 Motion

REVISED

Key objectives

By the end of this section, you should be able to:
● define speed and velocity, and recall and use the equations to calculate speed and average speed

● define acceleration and use the equation to calculate acceleration and know that a negative acceleration is a deceleration

- sketch, plot and understand the motion shown on distance–time and speed–time graphs
- determine from data or the shape of a distance–time graph or speed–time graph when an object is at rest, moving with constant speed, accelerating or decelerating
- calculate speed from the gradient of a distance–time graph and distance travelled from the area under a speed–time graph
- know the approximate value of the acceleration of freefall, *g*, for an object close to the Earth's surface

- determine from data or the shape of a speed–time graph when an object is moving with constant acceleration and changing acceleration
- calculate acceleration from the gradient of a speed–time graph
- describe the motion of an object falling in a uniform gravitational field with and without air or liquid resistance

Speed

Speed is defined as the distance travelled per unit time. **Velocity** is speed in a given direction. If someone sees a runner moving at 5 m/s in a northerly direction, then the runner's speed is 5 m/s and their velocity is 5 m/s north. The speed, *v*, can be calculated from the distance travelled, *s*, in a very short time, *t*, using the equation:

$$v = \frac{s}{t}$$

In most cases, speed is calculated using a much longer time. This is then the average speed of the object. The average speed is calculated using the equation:

$$\text{average speed} = \frac{\text{total distance travelled}}{\text{total time taken}}$$

Acceleration

Acceleration is the change in velocity per unit time. The acceleration, *a*, for a change in velocity, Δv, when the time taken for the change is Δt is given by:

$$a = \frac{\Delta v}{\Delta t}$$

A negative acceleration is called a **deceleration**.

Distance–time graphs

Distance–time graphs show how an object's distance changes with time. Figure 1.6 shows the motion of an object plotted on a distance–time graph.

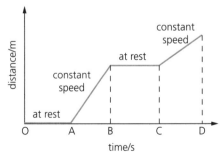

▲ **Figure 1.6 Distance–time graph**

The gradient of the graph for the section AB is greater than the gradient for section CD. This shows the object was moving at a faster constant speed at AB. The gradient of the distance–time graph is equal to the speed.

Skills

Calculating the gradient of a graph
To calculate the gradient of a graph, you need to read values for the change in y in a set change in x. See Figure 1.7.

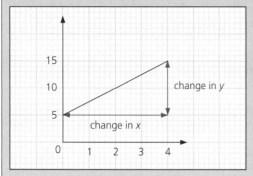

▲ **Figure 1.7 Calculating the gradient**

The gradient is then given by:

$$\text{gradient} = \frac{\text{change in } y}{\text{change in } x}$$

In Figure 1.7, the gradient = $\dfrac{(15 - 5)}{(4 - 0)}$ = 2.5.

If the speed increases, the object is accelerating. If the speed decreases, the object is decelerating. When the speed changes, the distance–time graph will curve. An upward curve shows the object is accelerating as the gradient is increasing. The solid green line in Figure 1.8 shows the object accelerating. A downward curve (the dashed green line in Figure 1.8) shows the object is decelerating.

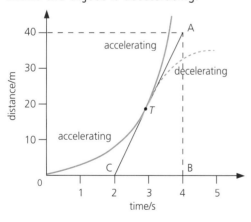

▲ **Figure 1.8 Non-constant speed**

The speed at any point on a distance–time graph where the object is changing speed is given by the gradient of the tangent drawn at that point. In Figure 1.8, the speed at time T is equal to the gradient of the tangent (line AC) drawn at that point.

$$\text{gradient} = \frac{40 - 0}{4 - 2} = \frac{40}{2} = 20 \text{ m/s}$$

speed at time T = 20 m/s

Determining the motion of an object from data

It is easy to interpret the motion from the shape of distance–time graphs, but you can also tell when you look at data in tables. When the object travels at a constant speed, the distance increases the same amount in equal times. When the object is stationary, the distance remains the same. When the distance increases in different amounts in equal times, the speed is changing. Table 1.1 shows how the distance of an object changes with time.

▼ **Table 1.1 Distance–time data**

Time/s	0	2	4	6	8	10	12	14
Distance/m	0	5	10	15	15	16	18	24
	Constant speed: Every 2 seconds distance increases by 5 m			Stationary: The distance remains the same		Changing speed: The distance travelled every 2 seconds is increasing		

Speed–time graphs

Speed–time graphs show the speed of an object over time. The area under the speed–time graph is the distance travelled (green shaded area in Figure 1.9).

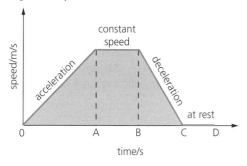

▲ **Figure 1.9 Speed–time graph showing acceleration, constant speed and deceleration**

In Figure 1.9, the object is accelerating between O and A, at constant speed between A and B and between B and C it is slowing down or decelerating.

The steeper the gradient of a speed–time graph, the greater the acceleration. In Figure 1.9 the deceleration is greater than the acceleration. The same change in speed happened in a much shorter time interval and the gradient is steeper.

Near the surface of the Earth the **acceleration of free fall** (g) is approximately constant and is equal to $9.8\,\text{m/s}^2$.

You can calculate the acceleration using the gradient of a speed–time graph. Figure 1.10 shows the speed–time graph for an object falling both without air resistance and with air resistance.

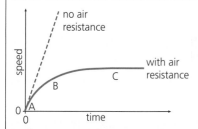

▲ **Figure 1.10 A body in free fall in the atmosphere**

Without air resistance the gradient of the graph is constant and equal to 9.8 m/s². However, with air resistance the acceleration decreases. You can see this because the gradient of the graph is decreasing. At point A in Figure 1.10, the speed is slow so there is negligible air resistance and the body has free fall acceleration. At point B, the speed is higher and there is some air resistance, so acceleration is less than free fall. At point C, the body has high speed and high air resistance, which is equal to its weight. Therefore, there is no acceleration – this constant speed is called the **terminal velocity**.

Sample question

REVISED

5   A runner completes an 800 m race in 2 min 30 s after completing the first lap of 400 m in 1 min 10 s. Find their average speed for the last 400 m.   [3]

Student's answer

Total time = 2 mins 30 s = (2 × 60) + 30 = 150 s

$speed = \dfrac{400}{150} = 2.67 \, m/s$ [2]

Teacher's comments

The student used the correct equation and the correct distance, but used the time for the whole race instead of the time for the last 400 m. The answer is quoted to 3 s.f.

Correct answer

Time = 2 min 30 s – 1 min 10 s = 1 min 20 s = 80 s   [1]

$speed = \dfrac{400}{80} = 5.0 \, m/s$ [2]

Sample question

REVISED

6   A car is moving in traffic and its motion is shown in Figure 1.11.

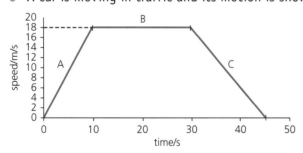

▲ **Figure 1.11**

   a   Choose from the following terms to describe the motion in parts A, B and C: acceleration, deceleration, steady speed.   [3]

   b   Calculate the total distance covered.   [5]

   c   Calculate the acceleration in part C.   [2]

Student's answers

a   Part A: acceleration; part B: deceleration; part C: steady speed   [1]

b   distance = speed × time = 18 × 45 = 810 m   [0]

c   acceleration $a = \dfrac{\Delta v}{\Delta t} = \dfrac{18}{15} = 1.2 \text{ m/s}^2$ [1]

Teacher's comments

a   The answers to parts B and C are the wrong way around.

b   The equation used is distance = average speed × time, but this is not appropriate, as the average speed is unknown. The student should have worked out the area under the graph, which equals the distance covered.

c   The calculation is correct but the student should have specified a negative acceleration.   [1 mark given]

Correct answers

a   Part A: acceleration; part B: steady speed; part C: deceleration   [3]

b   distance = area under graph [1]

Part A area = $\dfrac{1}{2}$ × 18 × 10 = 90 m [1]

Part B area = 18 × 20 = 360 m [1]

Part C area = $\dfrac{1}{2}$ × 18 × 15 = 135 m [1]

distance = total area = 90 + 360 + 135 = 585 m = 590 m to 2 s.f.   [1]

c   acceleration $a = \dfrac{\Delta v}{\Delta t} = \dfrac{-18}{15} = 1.2 \text{ m/s}^2$ [2]

Revision activity

Create a revision poster on motion. Start in the middle of sheet with the four types of motion you have to recognise: a) at rest, b) moving with a constant speed, c) accelerating, d) decelerating. Draw distance–time and speed–time graphs to represent each type of motion. Include all the key words used to describe motion and how motion can be calculated. Link these to your graphs. Swap posters and see how another student has summarised the same information.

Exam-style questions

Answers available at: www.hoddereducation.co.uk/cambridgeextras

5   Runner A runs 100 m in 20 seconds at a constant speed.
a   Sketch this information on a distance–time graph.   [3]
b   Calculate their average speed.   [2]
c   Runner B is twice as fast. Add a line to your distance–time graph and label it B. Assume they also run at a constant speed and run 100 m.   [2]

6   A bus accelerates at a constant rate from standstill to 15 m/s in 12 s. It continues at a constant speed of 15 m/s for 8 s.
a   Sketch this information on a speed–time graph.   [3]
b   Use the graph to calculate the total distance covered.   [2]
c   Calculate the average speed of the bus.   [2]
d   Calculate the acceleration in the first 12 seconds.   [2]

1.3 Mass and weight

Key objectives

By the end of this section, you should be able to:
- define mass and weight and know that weights (and masses) can be compared using a balance
- define gravitational field strength, g, and use the equation relating g, weight and mass
- describe how the weight of an object depends on the gravitational field it is in

Mass is the amount of matter in an object. The unit of mass is the kilogram, kg.

Weight is the gravitational force acting on an object that has mass. As it is a force, the unit of weight is the newton, N.

Weight, W, and mass, m, are related. The weight depends on the strength of the gravitational field the mass is in. **Gravitational field strength** is defined as the force acting per unit mass and is given by the equation:

$$g = \frac{W}{m}$$

Gravitational field strength has the same symbol g as the acceleration of free fall as they are equivalent. The units are different. Near the surface of the Earth, gravitational field strength is 9.8 N/kg and acceleration of free fall is 9.8 m/s².

A balance such as the one shown in Figure 1.12 compares an unknown weight with a known weight.

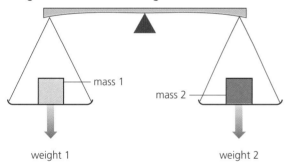

mass 1

mass 2

weight 1   weight 2

▲ **Figure 1.12 Balanced weights**

As mass determines weight, a balance also compares masses. In Figure 1.12, mass 1 = mass 2 because weight 1 = weight 2.

The mass of an object at rest is always the same as it depends on the matter in the object. However, the weight depends on the gravitational field the mass is in. A 1 kg mass has a weight of 9.8 N on Earth. Jupiter has a gravitational field strength of 25 N/kg. The same 1 kg mass would have a weight of 25 N on Jupiter.

Exam-style questions

Answers available at: www.hoddereducation.co.uk/cambridgeextras
7   A rover used to explore planets weighs 8820 N on Earth. On Mars the rover weighs 3330 N.
   a   Calculate the mass of the rover.   [2]
   b   Calculate the gravitational field strength on Mars.   [1]

Revision activity

Create flashcards for the definitions of the key terms in this section and the equation $g = \frac{W}{m}$.

1.4 Density

Key objectives

By the end of this section, you should be able to:
- define density, and recall and use the equation relating density, mass and volume
- describe how to determine the density of a liquid, a regularly shaped solid and an irregularly shaped solid, including appropriate calculations
- use density data to determine whether an object will float or sink in a liquid

  - use density data to determine whether one liquid will float on another liquid

Density is the mass per unit volume of a substance.

For a mass m with volume V the density ρ is given by the equation:

$$\rho = \frac{m}{V}$$

The units of density are kg/m^3.

Skills

Measuring the density of different substances

To find the density of a substance you must make accurate measurements of the mass and volume:

For a regularly shaped solid, measure the dimensions and work out the volume, then find the mass on a balance.

volume of a rectangular block = length × breadth × height

volume of a cylinder = $\pi r^2 h$

For an irregularly shaped solid, use a displacement method where the solid is placed in water (Figure 1.13). In method 1, the volume of the solid is the increase in the reading on the measuring cylinder (see Figure 1.13a). In method 2, where a displacement can is used, the volume of the solid is the volume of liquid displaced (see Figure 1.13b).

For a liquid, measure the volume in a measuring cylinder. To find the mass of the liquid, first find the mass of an empty beaker. Pour the liquid into the beaker and then find the total mass of the beaker and the liquid. Work out the mass of the liquid by subtraction of the mass of the beaker from the mass of the total.

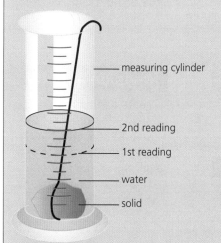

▲ **Figure 1.13a Measuring the volume of an irregular solid method 1**

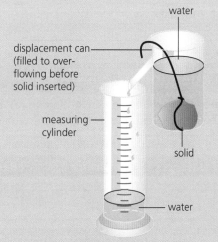

▲ **Figure 1.13b Measuring the volume of an irregular solid method 2**

Skills

Converting units

In your experiment, you probably measured the mass in grams and the volume in cm^3. This gives you a density in g/cm^3. To convert this to kg/m^3 you multiply by 1000.

For example, state which has the higher density: substance A at $0.8\,g/cm^3$ or substance B at $750\,kg/m^3$.

Both substances need to be in the same units of kg/m^3 so that you can compare them.

density of A = 0.8 × 1000 = $800\,kg/m^3$

Therefore, substance A has a greater density.

Floating and sinking

An object will sink in a liquid if it has density greater than the density of the liquid.

When two liquids do not mix, the liquid with the lower density will float on top of the liquid with higher density.

Sample question

7 The mass of an empty measuring cylinder is 185 g. When the measuring cylinder contains $400\,cm^3$ of a liquid, the total mass is 465 g. Find the density of the liquid. [4]

Student's answer

$$density = \frac{465}{400} = 1.16\,g/cm^3 = 1.2\,g/cm^3 \text{ to 2 s.f.}$$ [2]

Correct answer

mass of liquid = 465 − 185 = 280 g

$$density = \frac{280}{400} = 0.70\,g/cm^3$$ [4]

Teacher's comments

The student put the appropriate quantities into the correct equation and gave the correct units, but used the total mass instead of working out and using the mass of the liquid itself.

Exam-style questions

Answers available at: www.hoddereducation.co.uk/cambridgeextras

8 a Copy and complete the table by filling in the missing values. [3]

▼ **Table 1.2**

Substance	Mass/g	Volume/cm³	Density/g/cm³
A	540	200	
B	67.5		1.5
C		250	0.5

b State which of the substances would float in a liquid with a density of $1.2\,g/cm^3$. [1]

Revision activity

Create a mind map on density. Include how to calculate density, how to measure the density of a substance and how you use density to determine whether objects float.

9 A measuring cylinder containing $20\,cm^3$ of liquid is placed on a top-pan balance. The top-pan balance reads $150\,g$. More liquid is poured into the cylinder up to the $140\,cm^3$ mark and the top-pan balance now reads $246\,g$. A solid is gently lowered into the cylinder; the liquid rises to the $200\,cm^3$ mark and the top-pan balance reads $411\,g$. Calculate:
   a the density of the liquid   [3]
   b the density of the solid   [3]

10 A student has the same mass, $85\,g$, of two different liquids. Liquid A has a volume of $80\,cm^3$ and liquid B has a volume of $92\,cm^3$. Determine which liquid will float on top assuming the liquids do not mix.   [3]

1.5 Forces

1.5.1 Effects of forces

Key objectives

By the end of this section, you should be able to:
- know that forces may produce changes in the size and shape of an object
- describe an experiment to collect data for a load–extension graph and plot, sketch and understand the features of a load–extension graph

  - define the spring constant, and recall and use the equation and define the limit of proportionality

- determine the resultant force when two or more forces are acting along the same line
- understand that an object will remain at rest or continue at a constant speed in a straight line unless a resultant force acts on it
- understand that a resultant force may change the velocity of an object by changing its speed or direction

- recall and use the equation $F = ma$ to calculate the resultant force, F, and the acceleration, a, and know that the force and acceleration are in the same direction

- state how solid friction opposes motion between two surfaces and produces heating
- understand there is friction acting on an object as it moves through gas (air resistance) or a liquid (drag)

  - describe the motion of an object in a circular path and how the force is affected as the speed, radius of the circle and mass of the object change

Forces

Forces can change the size and shape of a body. You must be able to describe an experiment to measure the **extension** of an elastic solid, such as a spring, a piece of rubber or another object, with increasing load. The extension is the change in length of the object being stretched. For some materials, the load–extension graph is a straight-line graph through the origin. This means the load is directly proportional to the extension. This means doubling the force, doubles the extension. Not all load–extension graphs are linear, which means the force required to stretch the material changes as the material is stretched.

Spring constant

The **spring constant**, k, is defined as the force per unit extension. The units are N/m. The spring constant can be calculated using the equation:

$$k = \frac{F}{x}$$

For a linear load–extension graph, the spring constant stays the same. The spring constant will be the gradient of the graph. On a load–extension graph the **limit of proportionality** is the point at which the graph is no longer linear.

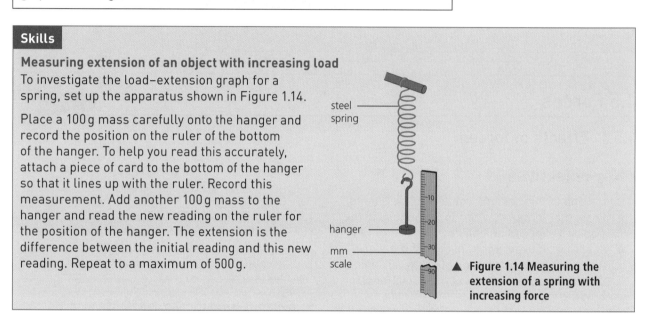
Forces and resultants

Force has both magnitude and direction. It is represented using an arrow to show the magnitude and direction the force acts. If more than one force acts on an object, you can find the resultant force. This is a single force which has exactly the same effect as all the forces added together. Figure 1.15 shows how to find the resultant of forces acting along the same line. If a question simply describes forces, it will help to sketch a force diagram showing the direction of each of the forces.

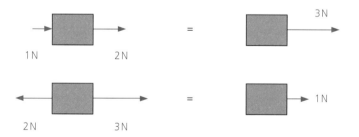

▲ Figure 1.15 Use addition or subtraction to find the resultant of forces acting in a straight line

If the resultant force acting on an object is zero, then the object will stay at rest or keep moving in a straight line at a constant speed.

If there is a resultant force acting on an object, then it changes velocity. This can mean a change in speed or/and a change in direction. Remember velocity is speed with direction.

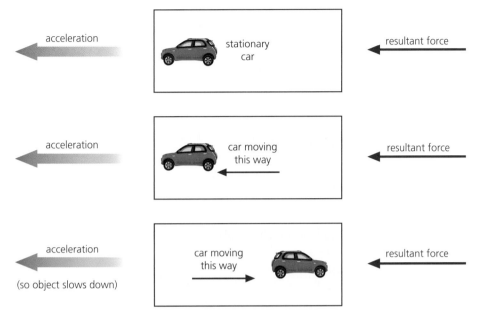

▲ **Figure 1.16 A resultant force changes the motion of an object**

Friction

Solid **friction** is a force that opposes one surface that is moving or trying to move over another. Friction results in heating. When an object moves through a gas or liquid, there is a friction force opposing the motion. This friction force in liquid is called drag and in air is called **air resistance**.

Relationship between resultant force and acceleration

You need to know and be able to use the equation $F = ma$, where F is the resultant force and a is the acceleration. The acceleration is in the direction of the resultant force.

When the resultant force is perpendicular to motion, the object follows a circular path. Some examples of this are shown in Table 1.3.

▼ **Table 1.3 Examples of circular motion**

Object	Force	Circular motion
Planet in orbit	Gravitational force towards the Sun	Planet moves around the Sun
Car turning a corner	Friction force	Car drives around the corner
Ball on a length of string	String tension	Ball moves around in a circle on the end of the string

Although the object may be moving at a constant speed, it is still accelerating as it is continually changing direction. This means the velocity is changing. Remember velocity is a vector.

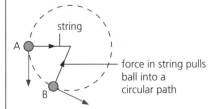

▲ **Figure 1.17 Diagram showing the direction of the force and velocity for a ball on a length of string**

The resultant force required to keep the object moving in the circle varies with the speed, radius and mass of the object:

● Increasing the speed, *increases* the force for the same mass and radius of circle.
● Increasing the radius, *decreases* the force for the same mass and speed.
● Increasing the mass, *increases* the force for the same speed and radius.

Sample question

REVISED ☐

8 An empty lift weighs 2000 N. Four people enter the lift and their total weight is 3000 N. After the button is pressed to move the lift, the tension in the cable pulling up from the top of the lift is 4000 N.
  a Work out the resultant force on the lift. [2]
  b State how the lift moves. [2]
  c Calculate the resultant acceleration ($g = 9.8$ N/kg). [3]

Student's answers

a Resultant force = 3000 + 2000 − 4000 = 1000 N [1]
b The lift will move down. [1]

c Mass of lift and people = $\dfrac{5000}{9.8}$ = 510.2 kg

Acceleration = $\dfrac{F}{m} = \dfrac{1000}{510.2}$ = 1.96 m/s² downwards [3]

Teacher's comments

a The student correctly worked out the size of the force but did not state the direction downwards.
b The words 'move down' are too vague.

c The student's answer is correct but has been quoted to 3 s.f.

Correct answers

a Resultant force = 3000 + 2000 − 4000 = 1000 N downwards [2]
b The lift will accelerate downwards. [2]

c Mass of lift and people = $\dfrac{5000}{9.8}$ = 510.2 kg

Acceleration = $\dfrac{F}{m} = \dfrac{1000}{510.2}$ = 2.0 m/s² downwards [3]

Exam-style questions

Answers available at: www.hoddereducation.co.uk/cambridgeextras

11 Figure 1.18 shows load–extension graphs for three different objects.

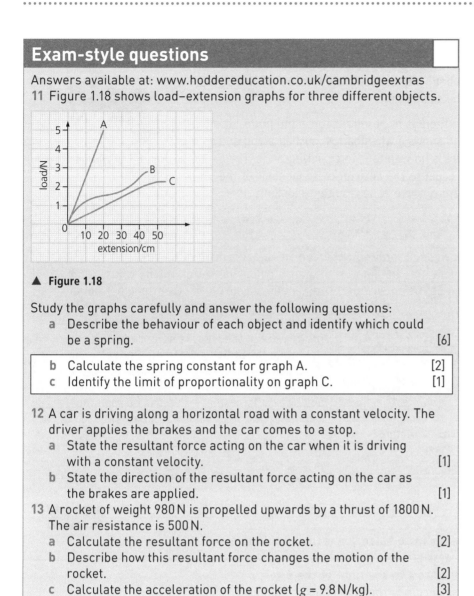

▲ **Figure 1.18**

Study the graphs carefully and answer the following questions:
-   a   Describe the behaviour of each object and identify which could be a spring. [6]
-   b   Calculate the spring constant for graph A. [2]
-   c   Identify the limit of proportionality on graph C. [1]

12 A car is driving along a horizontal road with a constant velocity. The driver applies the brakes and the car comes to a stop.
-   a   State the resultant force acting on the car when it is driving with a constant velocity. [1]
-   b   State the direction of the resultant force acting on the car as the brakes are applied. [1]

13 A rocket of weight 980 N is propelled upwards by a thrust of 1800 N. The air resistance is 500 N.
-   a   Calculate the resultant force on the rocket. [2]
-   b   Describe how this resultant force changes the motion of the rocket. [2]
-   c   Calculate the acceleration of the rocket (g = 9.8 N/kg). [3]

1.5.2 Turning effect of forces

Key objectives

By the end of this section, you should be able to:
-   describe and give examples of the turning effect or moment of a force
-   define moment and recall and use the equation for the moment of a force
-   apply the principle of moments and recall the conditions for an object in equilibrium
-   apply the principle of moments when there are more than two forces about a pivot

-   describe an experiment to show that an object in equilibrium has no resultant moment

Moment of a force

The **moment of a force** is a measure of its turning effect. Everyday examples of moments include spanners and the handle on a door. In each case, the effort is applied at a distance from the pivot to increase the turning effect.

A moment is defined by the equation:

$$moment = force \times perpendicular\ distance\ from\ pivot$$

Principle of moments

The **principle of moments** states when a body is in equilibrium, the sum of the clockwise moments about any point equals the sum of the anticlockwise moments about the same point.

▲ **Figure 1.19 Wheelbarrow in equilibrium**

An object is in **equilibrium** when there is no resultant force and no resultant moment on the object. Figure 1.19 shows a wheelbarrow pushed along at a constant velocity. Therefore, there is no resultant force acting on it. The clockwise moment of the load is equal to the anticlockwise moment of the upward force holding it up. Therefore, there is no resultant moment.

Skills

Demonstrating that there is no resultant moment when an object is in equilibrium

To demonstrate there is no resultant moment on an object in equilibrium, set up the apparatus shown in Figure 1.20.

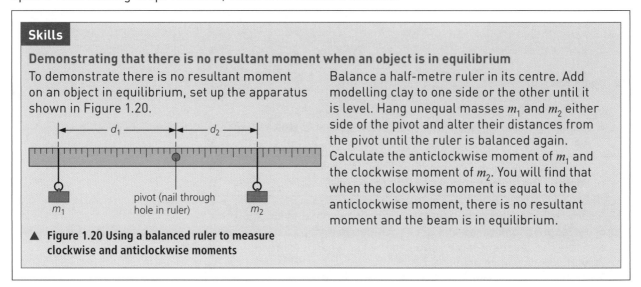

▲ **Figure 1.20 Using a balanced ruler to measure clockwise and anticlockwise moments**

Balance a half-metre ruler in its centre. Add modelling clay to one side or the other until it is level. Hang unequal masses m_1 and m_2 either side of the pivot and alter their distances from the pivot until the ruler is balanced again. Calculate the anticlockwise moment of m_1 and the clockwise moment of m_2. You will find that when the clockwise moment is equal to the anticlockwise moment, there is no resultant moment and the beam is in equilibrium.

Sample question

REVISED ☐

9 A student carries out an experiment to balance a regular 4 m long plank at its mid-point. A mass of 4 kg is placed 80 cm to the left of the pivot and a mass of 3.2 kg is placed 100 cm to the right of the pivot. Explain, *by calculating the moments,* whether the plank is balanced. Use $g = 10$ N/kg. [4]

▲ **Figure 1.21**

Student's answer

$4 \times 80 = 3.2 \times 100$, so the plank balances. [2]

Correct answer

anticlockwise moment = $40 \times 0.8 = 32$ N m [1]

clockwise moment = $32 \times 1 = 32$ N m [1]

anticlockwise moment = clockwise moment, so the plank balances. [2]

Teacher's comments

The student's calculation and conclusion are entirely correct, but the instruction in italic to calculate the moments was ignored.

Exam-style questions

Answers available at: www.hoddereducation.co.uk/cambridgeextras

14 A 450 N child sits 1.2 m from the pivot of a seesaw.
   a Calculate the moment of the child.   [1]
   b A second child sits 1.5 m from the pivot on the opposite side.
      The seesaw is in equilibrium. Calculate the weight of the
      second child.   [2]

15 A seesaw has a total length of 4 m and is pivoted in the middle.
   A child of weight 400 N sits 1.4 m from the pivot. A child of weight
   300 N sits 1.8 m from the pivot on the other side. A parent holds the
   end of the seesaw on the same side as the lighter child. Calculate
   the magnitude and direction of the force the parent must exert to
   hold the seesaw level.   [4]

Revision activity

Write four questions on moments with their solutions. Swap your questions with a partner and try each other's questions. Check the answers against each other.

1.5.3 Centre of gravity

Key objectives

By the end of this section, you should be able to:
● define centre of gravity and describe how its position relates to the stability of simple objects
● describe an experiment to find the centre of gravity of an irregularly shaped piece of card.

Centre of gravity

A body behaves as if its whole weight were concentrated at one point, called its **centre of gravity.** If you hang an object so it can swing freely, it will end up with its centre of gravity directly beneath the point of suspension. In a regular object of uniform shape and density, the centre of gravity will be in the geometric centre.

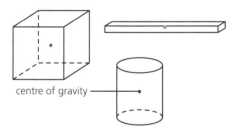

centre of gravity

▲ **Figure 1.22 Centre of gravity is in the geometric centre of uniform shape and density objects**

Skills

Finding the centre of gravity

To find the centre of gravity of an irregularly shaped plane lamina (thin sheet) of cardboard, simply make a hole in the card. Use the hole to hang the card from a nail held in a clamp stand. The card must be able to swing freely so that it comes to rest with the centre of gravity directly below the nail. Tie a plumb line to the nail and mark its position on the card AB as shown in Figure 1.23. Make a second hole in the card and repeat the procedure making the line CD. Where the lines cross is the centre of gravity.

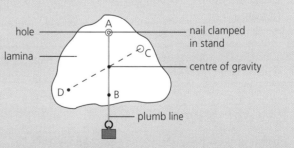

▲ **Figure 1.23 How to find the centre of gravity of an irregularly shaped lamina**

Toppling

The position of the centre of gravity affects the stability of an object. If an object is pushed, it will topple if the vertical line from the centre of gravity falls outside the base as in Figure 1.24a. It will not topple if the vertical line stays within the base as in Figure 1.24b.

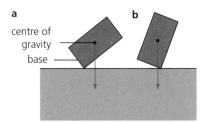

▲ **Figure 1.24 An object will topple if the vertical line from its centre of gravity falls outside the base. a The object topples and b the object will remain standing**

To increase the stability of an object:

- lower the centre of gravity
- increase the area of its base

Sample question

10 Explain why the model parrot will only stay on its perch if the bulldog clip is in place. [2]

Student's answer

The bulldog clip lowers the centre of gravity. [1]

Teacher's comments

The centre of gravity is lowered but the student did not mention its position relative to the perch.

card

perch

bulldog clip

▲ **Figure 1.25**

Correct answer

The bulldog clip moves the centre of gravity to directly below the perch, so the parrot is stable. [2]

Exam-style questions

Answers available at: www.hoddereducation.co.uk/cambridgeextras
16 Describe how a Bunsen burner is designed so that it is very hard to topple. [2]

1.6 Momentum

Key objectives

By the end of this section, you should be able to:
- define momentum, impulse and resultant force, and recall and use the correct equations to calculate them
- apply the principle of conservation of momentum to solve simple problems

Momentum (p) is the product of mass (m) and velocity (v):

$$p = mv$$

The units of momentum are kg m/s. Momentum is a vector quantity and so the direction is important.

Conservation of momentum

In any interaction between bodies, the total momentum is conserved. This is known as the **principle of conservation of momentum**. This includes explosions in rockets as well as collisions. In an explosion such as a cannon firing, the total momentum before firing is zero. After firing, the cannonball moves forward and the cannon rolls backwards. Their momentum is equal and opposite.

Force and momentum

The **impulse** of a force is defined as the product of the force (F) and the time over which the force acts (Δt).

$$\text{impulse} = F\Delta t$$

In any interaction, the impulse exerted on a body = change of momentum.

$$F\Delta t = \Delta(mv) \qquad \text{or} \qquad F\Delta t = \Delta p$$

Earlier in this section, you used the equation $F = ma$ for resultant force. This relationship gives you another equation for resultant force and a new definition. **Resultant force** is the change in momentum per unit time:

$$F = \frac{\Delta p}{\Delta t}$$

Sample question

11 A truck of mass 1800 kg moving with a velocity of 4 m/s to the right collides with a truck of mass 1200 kg moving with a velocity of 1 m/s to the left. The truck of mass 1800 kg has a velocity of 1.5 m/s to the right after the collision. Find the final velocity of the 1200 kg truck. [4]

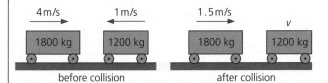

before collision   after collision

▲ **Figure 1.26**

Student's answer

momentum before collision = $1800 \times 4 + 1200 \times 1 = 7200 + 1200$
$= 8400$ kg m/s   [0]

momentum after collision = $1800 \times 1.5 + 1200v = 2700 + 1200v$ kg m/s   [1]

momentum is conserved: $2700 + 1200v = 8400$ [1]

$1200v = 5700$

$v = \dfrac{5700}{1200} = 4.75$ m/s $= 4.8$ m/s to 2 s.f.   [1]

Teacher's comments

The student made a good attempt at the question. The working was well set out. However, the student did not realise that direction is significant, as momentum is a vector quantity.

Correct answer

Consider that the positive direction is to the right and assume that v is also to the right.

momentum before collision = $(1800 \times 4) - (1200 \times 1) = 7200 - 1200$
$$= 6000 \, kg \, m/s \quad [1]$$

momentum after collision = $(1800 \times 1.5) + 1200v = 2700 + 1200v \, kg \, m/s$ [1]

momentum is conserved: $2700 + 1200v = 6000$ [1]

$1200v = 3300$
$$v = \frac{3300}{1200} = 2.75 \, m/s = 2.8 \, m/s \text{ to 2 s.f. to the right} \quad [1]$$

Exam-style questions

Answers available at: www.hoddereducation.co.uk/cambridgeextras

17 A railway truck of mass 6000 kg moving at 6 m/s collides with a truck of mass 10 000 kg moving at 2 m/s in the same direction. The two trucks couple and move on together at 3.5 m/s.
   a Carry out a calculation to confirm that momentum is conserved. [2]
   b Determine whether kinetic energy is conserved in the collision. [3]
   c Comment on your answer to part b. [1]
18 During a kick, a 450 g football accelerates from 0 to 25 m/s. The foot and the ball are in contact for 0.02 s. Calculate the force on the ball. [2]

Revision activity

Write the words 'force', 'velocity', 'momentum' and 'time' on a blank page. Try to link the words with an explanation, e.g. link momentum and velocity and write 'momentum = mass × velocity' on the linking line. See how many links you can make. You can add other words from this topic so far. Compare your work with a partner and see if there are any links you missed.

1.7 Energy, work and power

REVISED

1.7.1 Energy

Key objectives

By the end of this section, you should be able to:
- recall the different energy stores and describe how energy is transferred between these stores
- know the principle of conservation of energy, and apply and interpret simple energy flow diagrams

- recall and use the equations to calculate kinetic energy and the change in gravitational potential energy
- interpret complex energy flow diagrams including Sankey diagrams

Energy stores

Energy may be stored in many different ways.

▼ Table 1.4 Energy stores and their descriptions

Stores of energy	Description	Equations
Kinetic energy	Energy due to motion, e.g. a car moving, a stone falling, a person running	$E_k = \frac{1}{2}mv^2$
Gravitational potential energy	Energy due to position, e.g. any object above the Earth's surface such as a book on a high shelf or water in a mountain lake	$\Delta E_p = mg\Delta h$
Chemical energy	Food, fuel and batteries are stores of chemical energy. The energy is released by chemical reactions.	
Elastic (strain) energy	Energy stored due to the stretching or bending of materials, e.g. stretching a rubber band, compressing or extending a spring	

Stores of energy	Description	Equations
Nuclear energy	The energy stored in the nucleus of an atom. It can be transferred by nuclear reactions such as fission in nuclear reactors or fusion in the Sun	
Electrostatic energy	Energy stored by charged objects	
Internal energy	Also called thermal energy	

Δ is the Greek letter delta. It is used to represent change in a variable, in this case, the change in the gravitational potential store with a change in height.

Energy transfers

The **principle of conservation of energy** states that energy cannot be created or destroyed. However, energy can be transferred between stores by:

● forces (mechanical working)

● electrical currents (electrical working)

● heating through conduction, convection and radiation (Topic 2.3)

● waves (electromagnetic, sound and other waves)

You can represent these energy transfers using a simple energy flow diagram such as in Figure 1.27.

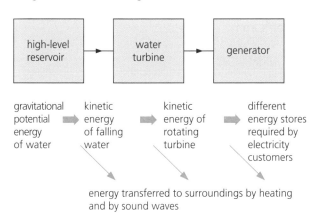

▲ **Figure 1.27 Energy transfers in a hydroelectric power scheme**

At each stage of energy transfer, some energy is transferred less usefully. As you can see in Figure 1.27, not all the energy in the gravitational potential store is transferred to the energy stores required by electricity customers. Some of the energy is transferred to the surroundings.

In a Sankey diagram, the thickness of the bars represents the amount of energy transferred at each stage. This is useful to see the proportion of energy usefully transferred at each stage in a process. As you can see in Figure 1.28, only 30% of the energy input is transferred to the energy stores required by electricity customers.

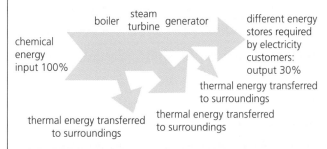

▲ **Figure 1.28 Energy transfers in a hydroelectric power scheme**

> **Conservation of energy**
>
> As an object falls it loses gravitational potential energy and gains kinetic energy. If it falls from rest and you ignore air resistance, the kinetic energy on reaching the ground is equal to the loss in gravitational potential energy:
>
> $$E_k = \Delta E_p \text{ or } \frac{1}{2}mv^2 = mg\Delta h$$
>
> This is also true if you throw a ball vertically upwards in the air. As the ball will stop for a moment when it reaches its maximum height, the gain in gravitational potential energy is equal to the loss in kinetic energy.

Sample question

REVISED ☐

▲ **Figure 1.29**

12 A person winds up the spring of the clockwork radio shown in Figure 1.29 using the muscles in their hand and arm. The internal spring then unwinds to provide energy to power the radio.

  a Describe the energy transfer between the muscles in the person's arm and the spring. [3]

  b Describe the process by which energy is transferred by the circuits in the radio. [1]

  c Name two ways energy is transferred from the radio. [2]

Student's answers

> a Chemical energy is stored in their muscles and it is stored as elastic energy in the spring. [2]
> b Electrical working. [1]
> c The energy transfer is radio and sound waves. [1]

Teacher's comments

> a The student has identified both stores of energy but not how the energy was transferred. In this case. by mechanical working.
> b Correct answer.
> c Energy is transferred by sound waves but radio waves transfer energy to the radio and not from.

Correct answers

> a Chemical energy stored in their muscles is transferred to elastic energy in the spring by mechanical working. [3]
> b Electrical working. [1]
> c Energy is transferred from the radio by sound waves and by heating. [2]

Revision activity

Make flashcards for each energy store and type of energy transfer. Use the cards to create energy transfer diagrams for different processes, e.g. a wind turbine, a solar cell, dropping a ball, throwing a ball, a pendulum, a torch. Compare your diagrams with a partner.

1.7.2 Work

Key objectives

By the end of this section, you should be able to:
- understand that mechanical work or electrical work done is equal to the energy transferred
- recall and use the equation to calculate mechanical work

In any energy transfer, **work** is done. Mechanical work is done when a force moves though a distance. The greater the force, the more work is done. The greater the distance moved, the more work is done.

work done = force × distance moved in the direction of the force

Know and be able to use the following equation:

$W = Fd = \Delta E$

where W is work done and ΔE is energy transferred.

To calculate work, you must identify the force and the distance moved in the direction of the force. If you walk up the stairs, you transfer energy to the gravitational potential store by mechanical work. The force is your weight ($W = mg$) and the distance is the vertical height of the stairs.

1.7.3 Energy resources

Key objectives

By the end of this section, you should be able to:
- describe the different ways useful energy is obtained or electrical power is generated, and the advantages and disadvantages of each method
- understand what is meant by the efficiency of energy transfer

- understand that the Sun releases energy through nuclear fusion and this is the main source of energy for most of the energy resources
- know that research is being carried out to find out if it is possible to have large-scale nuclear fusion to produce electricity
- define efficiency, and recall and use equations to calculate efficiency

Energy resources

There are many different energy resources. They can either be:

- **non-renewable** – cannot be replaced when used up
- **renewable** – can be replaced which means they cannot be used up

When choosing an energy resource, you have to consider its availability, reliability, scale and environmental impact. Most schemes involve a

generator. When a generator is turned it produces electricity (Topic 4.5).
Table 1.5 summarises this for the different energy resources.

▼ **Table 1.5 Energy resources**

Source	How the useful energy is obtained or electrical power generated	Renewable (R) or non-renewable (NR)	Availability	Reliable	Possible scale	Environmental impact
Fossil fuels	Chemical energy store in the fuels is released when they are burnt. This heats water in a boiler, making steam. The steam turns a turbine, which in turn drives a generator.	NR	Oil and gas running low; coal for the next 200 years	Yes	Large	Burning produces carbon dioxide (causes global warming) and sulfur dioxide (causes acid rain).
Water	Hydroelectric–gravitational potential energy in the water behind the dam. As the water flows through the dam, turbines are turned, which in turn drive generators.	R	Only some areas of the world have suitable sites	Yes	Medium	Loss of habitat for wildlife; land used for farming or forestry may be lost due to flooding for the dam.
	Tidal – same principle as hydroelectric	R	Only some areas of the world have suitable sites	Yes	Small	Destroys habitats for wildlife and causes problems in shipping routes
	Waves – energy of the waves is used to drive a generator.	R	Useful for island communities	Yes	Small	Problems to shipping
Geothermal	Cold water is pumped into hot rocks below the Earth's surface. The steam is used to turn a turbine, which then drives a generator.	R	Only certain parts of the world have rocks near enough to the surface that are hot enough for this to work.	Yes	Medium	Some (open-loop) designs have air emissions, although at a much lower level than burning fossil fuels: carbon dioxide (causes global warming) and hydrogen sulfide which changes to sulfur dioxide (causes acid rain). However, closed-loop designs do not.
Nuclear	Nuclear fission releases energy which is used to generate steam, which turn turbines which drive a generator.	NR	Available to countries with nuclear power stations	Yes	Large	Radiation; radioactive waste some of which has to be stored for thousands of years (Topic 5.2)
Radiation from the Sun	Solar cells – light used to generate electricity	R	Anywhere	No	Small	Large areas of solar cells to generate more electricity cover areas of land which could be used for food production.

Source	How the useful energy is obtained or electrical power generated	Renewable (R) or non-renewable (NR)	Availability	Reliable	Possible scale	Environmental impact
	Solar panels – infrared radiation heats water directly. Solar furnace – Using mirrors to focus the energy on a boiler, producing steam which turns a turbine which drives a generator.	R	Anywhere – but more effective closer to the equator	No	Small	
Wind	Turbine turned directly by the wind.	R	Coastal and upland sites best	No	Medium and small	Noisy; wind farms at sea can cause problems for shipping. Hazard to migrating birds.
Biomass	Chemical energy store in the biofuels.	R	Anywhere		Small	Produces carbon dioxide but is carbon neutral as carbon dioxide absorbed as biomass grows. Land used for food production may be lost to biofuel growth.

The Sun is the source of all energy resources except geothermal, nuclear and tidal. Energy is released by nuclear fusion in the Sun (Topic 5.1). Research is being done to try to reproduce fusion on Earth at a large scale. Currently it is not possible.

Efficiency

As you know, when energy is transferred form one store to another it is not all usefully transferred. How much energy is transferred usefully is described as the **efficiency**. In an efficient energy transfer, more of the energy input becomes useful energy output. A coal-fired power station has an efficiency of about 30%. This means that for every 100 J of energy released from burning coal only 30 J will be transferred to the electricity customers. However, a wind turbine at peak wind can have an efficiency of 50%. More of the energy has been usefully transferred.

Calculating efficiency

The **efficiency** of a device is the percentage of the energy supplied to it that is usefully transferred. You can calculate the efficiency using

$$\text{efficiency} = \frac{\text{useful energy output}}{\text{total energy input}} \times 100\%$$

or

$$\text{efficiency} = \frac{\text{useful power output}}{\text{total power input}} \times 100\%$$

Efficiency can never be more than 100%. This is because of the principle of conservation of energy.

Skills

Expressing efficiency

Efficiency can be expressed as a percentage or as a fraction written as a decimal number. To convert from a percentage to a fraction, simply divide the percentage efficiency by 100. For example, 45% is 0.45. As efficiency is always less than 100%, the number is always less than 1.

For example, a wind turbine is described as having an efficiency of 20%. Calculate the total energy input if the electrical energy transferred from the turbine is 300 J.

$$\text{efficiency as a decimal number} = \frac{20}{100} = 0.2$$

$$\text{efficiency} = \frac{\text{useful energy output}}{\text{total energy input}}$$

$$0.2 = \frac{300}{\text{total energy input}}$$

$$\text{total energy input} = \frac{300}{0.2} = 1500 \text{ J}$$

Sample question

REVISED ☐

13 For each of the following two statements, give one strength and one weakness and write a conclusion.

  a A supporter of nuclear power states that it should be more widely used as there is no pollution. [3]

  b A supporter of coal-fired power stations states that nuclear power plants cannot be controlled and might explode like atomic weapons. [3]

Student's answers

  a Nuclear power stations are constantly leaking radiation, which heats up the atmosphere. [0]

  b Nuclear power stations might be damaged in an earthquake or attacked by terrorists. This means there is a constant risk of a serious release of radioactive material. [2]

Correct answers

Possible answers could be:

  a Nuclear power stations produce no carbon dioxide or other air pollution, which is a powerful argument in their favour. Nuclear power stations produce waste, some highly radioactive, which is very hard to dispose of, as it has a half-life of many centuries. This is a powerful argument against nuclear power. Because of concerns about the production of greenhouse gases and global warming, the lack of air pollution is a substantial benefit. The issue of radioactive waste is a serious problem. The overall judgment is a matter of opinion with different countries reaching different decisions. [3]

  b There have been examples of nuclear power plants overheating causing meltdown of the radioactive core, which has caused considerable release of radioactive substances into the environment. This happened in the Chernobyl incident as a result of an unauthorised test and in Japan following tsunami damage. However, the statement exaggerates the dangers, as the nuclear material is arranged in a manner that could not cause an explosion like that of a nuclear weapon. In addition, as experience is gained, the likelihood of future incidents decreases. Again, the two arguments have to be weighed against each other and the overall judgement is a matter of opinion with different countries reaching different decisions. [3]

Teacher's comments

  a The student has made *no* real attempt to address the issues or draw a conclusion. If there are three marks for a question, try to make three points.

  b This answer is incomplete, but the student has made two relevant points and made a basic conclusion.

Revision activity

Draw a table to summarise the advantages and disadvantages of each energy resource.

1.7.4 Power

Power is the work done per unit time and the energy transferred per unit time. Remember work done is a measure of the energy transfer. The unit of power is the watt, W.

$$\text{power} = \frac{\text{work done}}{\text{time taken}} \quad \text{or} \quad \text{power} = \frac{\text{energy transferred}}{\text{time taken}}$$

$$P = \frac{W}{t} \quad \text{or} \quad P = \frac{\Delta E}{t}$$

Sample question

14 The two cranes shown in Figure 1.30 are lifting loads at a port. Crane A raises a load of 1000 N to a height of 12 m in 10 s. Crane B raises the same load of 1000 N to the same height of 12 m but takes 12 s.

▲ **Figure 1.30**

  a Compare the work done by the two cranes.   [2]
  b Compare the power of the two cranes.   [2]
  c Calculate the energy transferred and the power of each crane.   [4]

Student's answers

  a *Both cranes do the same amount of work because the force and distance moved are the same.*   [2]
  b *Crane B has more power because the amount of work done is the same but the time is greater.*   [0]
  c *energy transferred by each crane = 1000 × 12 = 12 000*
    power of A = 12 000 × 10 = 120 000
    power of B = 12 000 × 12 = 144 000 = 140 000 to 2 s.f.   [1]

Teacher's comments

a Correct answer.
b The student has confused the relationship; the shorter the time taken, the greater the power.
c The calculation of energy transferred is correct, except that the unit (J) has been omitted. Both power calculations are incorrect because the wrong equation has been used; the unit of power (W) has also been omitted.

Correct answers

  a Both cranes do the same amount of work because the force and distance moved are the same.   [2]
  b Crane A has more power because the amount of work done is the same but less time is taken.   [2]
  c Energy transferred by each crane = 1000 × 12 = 12 000 J

$$\text{Power of A} = \frac{12\,000}{10} = 1200\,\text{W}$$

$$\text{Power of B} = \frac{12\,000}{12} = 1000\,\text{W} \qquad [4]$$

Exam-style questions

Answers available at: www.hoddereducation.co.uk/cambridgeextras

19 Copy and complete the table below to show the energy transfers in different devices. [5]

Device	Energy store at the start that decreases	Energy transfer process	Energy store at the end that increases
battery-powered fan	chemical energy		kinetic energy
roller coaster			kinetic energy
catapult	elastic energy		

20 A bungee jumper of mass 60 kg jumps from a bridge tied to an elastic rope that becomes taut after they fall 10 m. Consider the jumper when they have fallen another 10 m and are travelling at 15 m/s.
   a State a store of energy that has decreased. [1]
   b State two stores of energy that have increased. [2]

   c Calculate the change in gravitational potential energy of the bungee jumper when they have fallen 20 m. [2]
   d Calculate how much energy is stored in the rope. Ignore air resistance. [3]
21 Calculate the maximum height reached by a 60 g ball thrown vertically up in the air. The speed of the ball when it left the hands is 12 m/s. [3]

22 A 15 kg box is moved. In each case, calculate how much work is done.
   a The box is dragged along the floor with a force of 25 N for 2 m. [1]
   b The box is lifted onto a shelf 1.5 m high. [2]
23 Supporters and opponents are discussing a proposed new wind farm of 20 large wind turbines. The supporters say that the wind farm will use energy from a renewable source, not pollute and provide reliable energy. The opponents admit it will use energy from a renewable source, but say that it will not be reliable and it will pollute.
   a Comment on the arguments of the supporters and the opponents. [4]
   b Is it correct to say 'that the wind farm will use energy from a renewable source'? [2]
   c Write down one other source of renewable energy and one source of non-renewable energy. [2]

24 Calculate the efficiency of a solar cell if the power input from the Sun is 1080 W and the power output is 432 W. [2]
25 A motor lifts a load of 40 N a distance of 1.4 m. The energy input to the motor is 70 J. Calculate the efficiency of the motor. [3]

26 a Calculate the power if a microwave transfers 96 000 J in 2 minutes. [2]
   b Calculate the average power if a 750 N person climbs stairs 15 m high in 30 s. [2]

27 In a small-scale hydroelectric power scheme, 24 kg of water falls every second through a vertical height of 60 m from the reservoir to the turbine. The electrical output is 11 kW. Calculate the efficiency of the scheme. [3]

Revision activity

Write five questions on power and their answers. Include at least one where you have to rearrange the equation. Swap questions with a partner and check you agree with the solutions.

1.8 Pressure

REVISED

Key objectives

By the end of this section, you should be able to:
- define pressure, and recall and calculate pressure using the correct equation
- describe how pressure varies with force and area in everyday examples
- describe how pressure in a liquid varies with depth and density of the liquid

- recall and use the equation to calculate the change in pressure in a liquid

Pressure

Pressure is defined as the force per unit area. To calculate the pressure, you need to use the following equation:

$$p = \frac{F}{A}$$

The unit of pressure is the pascal (Pa). A force of 1 N on an area of 1 m² exerts a pressure of 1 Pa.

The greater the area in contact as a force is applied, the less the pressure. For example, snowshoes and skis have a large surface area to stop the person wearing them sinking into the snow. They have the same weight as a person wearing normal shoes, but the pressure is less. A nail is designed with a small area of contact so that there is a high pressure when a force is applied. This allows the nail to be hammered into the wood.

Liquid pressure

Pressure beneath a liquid surface depends on the depth and the density of the liquid.

- The greater the depth in a given liquid, the greater the pressure. This is because as you increase in depth there is a greater weight of liquid above you. Figure 1.31a shows how the pressure is greater at the bottom of the column of liquid. Figure 1.31b shows that at one depth the pressure acts equally in all directions.

- At a given depth, the greater the density of the liquid, the greater the pressure. This is because a higher density liquid has a greater weight per unit volume. Remember the density is the mass per unit volume.

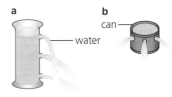

▲ **Figure 1.31 Pressure in a liquid**

You should know and be able to use the following equation:

$$\Delta p = \rho g \Delta h$$

where Δp is the change in pressure between the surface and that depth, ρ is the density of the liquid, g is the acceleration due to gravity and Δh is the depth below the surface of the liquid.

Skills

How many significant figures to use for the final answer

Remember only the final answer in the calculation should be rounded. Use the number in the calculator as you go through a calculation with more than one stage. The final answer should be the same number of significant figures (s.f.) as the number of the significant figures in the values used to calculate, e.g. if you know the depth is 5.00 m and the density is 1020 kg/m³, then you know the depth and density to 3 significant figures. However, if you use $g = 9.8$ m/s², you only know the value to 2 significant figures. This means your final answer should be quoted to 2 significant figures.

For example, calculate the pressure exerted on the floor by a 12 kg sack with an area of 0.015 m².

$W = mg = 12 \times 9.8 = 117.6\,\text{N}$

$$\text{pressure} = \frac{F}{A} = \frac{117.6}{0.015} = 7840\,\text{N} = 7800\,\text{N to 2 s.f.}$$

(All values used only given to 2 s.f.)

Sample question

15 Some students are playing a ball game in the sea and the ball is pushed 60 cm below the surface of the water
(Density of seawater = 1.025 × density of freshwater.)
   a Explain how the pressure on the ball at a depth of 60 cm below the surface of the sea compares with the pressure just below the surface. [2]
   b Explain how the pressure on the ball at a depth of 60 cm below the surface of the sea compares with the pressure 60 cm below the surface of a freshwater lake. [2]

   c Calculate the pressure on a point on the ball 60 cm below the surface of the sea (density of freshwater = 1000 kg/m³). [2]

Student's answers

   a The pressure increases. [1]
   b The pressure on the ball below the surface of the sea is greater because seawater has a greater density. [1]

   c pressure = $\Delta h \rho g = 0.6 \times 1000 \times 9.8 = 5900\,\text{Pa}$ to 2 s.f. [1]

Correct answers

   a The pressure increases because the ball is at a greater depth in the same liquid. [2]
   b The pressure on the ball below the surface of the sea is greater because seawater has a greater density and both balls are at the same depth. [2]

   c density of sea water = 1.025 × 1000 kg/m³ = 1025 kg/m³
   $\Delta p = \Delta h \rho g = 0.6 \times 1025 \times 9.8 = 6027\,\text{Pa} = 6000\,\text{Pa}$ to 2 s.f. [2]

Teacher's comments

   a The statement is correct but there is no explanation.
   b The statement is correct and the reason is also correct, but not quite complete. The student should have mentioned that the comparison was at the same depth below each surface.

   c The pressure has been calculated in the correct way but at a depth of 60 cm below the surface of freshwater instead of seawater.

Exam-style questions

Answers available at: www.hoddereducation.co.uk/cambridgeextras
28 Describe how a drawing pin is designed so that it can be pushed into a notice board without hurting you. [2]
29 A block of marble weighing 4900 N has a base in contact with the ground of 0.80 m by 1.30 m. Calculate the pressure on the floor. [2]
30 Dams are built across rivers to trap water behind. They are built so that they are much thicker at the bottom than the top.
   a Explain why they are thicker at the bottom than the top. [2]

   b The depth of water behind a dam is 78.0 m. Calculate the pressure at the bottom of a dam. Density of water = 1000 kg/m³. [1]

Revision activity

Create a mind map on pressure. Include any equations and some everyday examples of low- or high-pressure situations

2 Thermal physics

Key terms

Term	Definition
Absolute zero	Lowest possible temperature: −273°C or 0K
Condensation	Change of a gas to a liquid
Conduction	Flow of thermal energy transferred through matter from places of high temperature to places of low temperature without movement of matter as a whole
Convection	Flow of thermal energy through a fluid from places of high temperature to places of low temperature by movement of the fluid itself because of change of density
Degrees Celsius	°C; unit of temperature
Kelvin	K; SI unit of temperature; a kelvin has same size as a degree Celsius but 0°C = 273K
Molecule	Combination of atoms
Particle	Any small piece of a substance; it could be one ion, electron, atom or molecule or billions of them
Radiation of thermal energy	Transfer of thermal energy from one place to another by infrared electromagnetic waves
Temperature	A measure of the average kinetic energy of the molecules of a body
Thermal energy	Energy of the molecules of a body
Vaporisation	Change of a liquid to a gas
Specific heat capacity	Energy needed per unit mass per unit temperature rise

2.1 Kinetic particle model of matter

2.1.1 States of matter

> **Key objectives**
>
> By the end of this section, you should be able to:
> ● know the properties of solids, liquids and gases and the terms for the changes in state between them

Solids have a definite shape and size and are hard to compress.

Liquids have a definite volume but adopt the shape of their container. They are easier to compress than solids but still not easily compressed.

Gases have no definite size or shape but fill their container and adopt its shape.

Changes of state
Melting occurs when a solid becomes a liquid.

Solidification or freezing occurs when a liquid becomes a solid.

Evaporation or boiling occurs when a liquid becomes a gas.

Condensation occurs when a gas becomes a liquid.

2.1.2 Particle model

Key objectives

By the end of this section, you should be able to:
- explain how the kinetic particle model explains the nature of solids, liquids and gases

  - know that the forces and distances between particles and the particle motion affect the properties of solids, liquids and gases

- describe the relationship between particle motion and temperature and understand the concept of absolute zero

- describe pressure and change in pressure in terms of particle motion and their collisions with a surface

  - describe pressure and changes in pressure in terms of force per unit area
  - distinguish between microscopic particles, atoms and molecules

- describe and explain Brownian motion

All matter is made up of **particles** (atoms, **molecules**, ions, electrons) in motion. The higher the temperature, the faster the motion of the particles. Almost always, matter expands with increases in temperature. Particles have their least kinetic energy at absolute zero, the lowest possible temperature.

Solids

Key features of solids:

- Particles are close together.
- Particles vibrate about fixed points in a regular array or lattice.
- The rigid structure of solids results from these fixed positions.
- As temperature increases, the particles vibrate further and faster. This pushes the fixed points further apart and the solid expands.

- There is only a very slight expansion of a solid with increases in temperature, e.g. the length of an iron rod increases by about 0.1% when it is heated from 20°C to 100°C.
- The positions of particles in a solid are fixed because the attractive and repulsive forces between neighbouring particles are balanced.
- The strong attractive forces between particles of a solid give them a rigid structure. The weaker attractive forces between particles of a liquid hold the liquid together but without a rigid structure. The lack of attractive forces between particles of a gas allow them to move freely within their container.

Skills

Drawing simple particle diagrams
Draw a diagram to show the arrangement and motion of particles of a solid. Use the model of Figure 2.1 to guide you.
- Draw the particles as rows of circles regularly arranged.
- Add two-way arrows to show vibration.
- Label the arrows 'vibration of the particles in the lattice'.

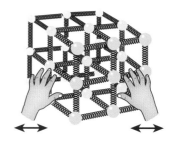

▲ Figure 2.1 A model of the behaviour the particles of a solid

Liquids

Key features of liquids:

- Particles are slightly further apart than in solids.
- Particles are still close enough to keep a definite volume.
- The main motion of the particles is vibration. The particles also move randomly in all directions, not being fixed to each other.
- As temperature increases, the particles move faster and further apart, so the liquid expands. One exception to this is that, when liquid water is heated from 0°C to 4°C, its structure changes, so it contracts instead of expands.

- The forces between particles are too weak to keep them in a definite pattern but are enough to hold them to the bulk of the liquid.
- There is a small expansion of a liquid with increases in temperature, e.g. the volume of many liquids increases by about 4% when heated from 20°C to 100°C.

Skills

Drawing simple particle diagrams

Draw a diagram to show the arrangement and motion of particles of a liquid. Use the model of Figure 2.2 to guide you.

- Draw some of the particles as rows of circles regularly arranged.
- Add other circles separated from the regular arrangement.
- Draw two-way arrows to show vibration of the particles regularly arranged.
- Label the arrows 'most particles vibrate'.
- Add labels to separated particles 'some particles are free to move through the liquid'.

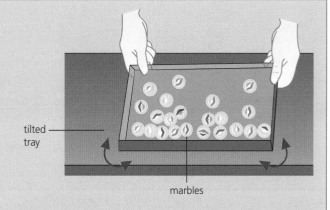

tilted tray

marbles

▲ **Figure 2.2 A model of the behaviour of the particles of a liquid**

Gases

Key features of gases:

- Particles are much further apart than in solids or liquids.
- Particles move much faster than in solids or liquids.
- There is no definite volume. Particles move throughout the available space.
- Particles constantly collide with each other and the container walls.

- There are no forces between particles except during a collision.

- There are no forces between particles except during a collision.
- Gases have low densities.
- The higher the temperature, the faster the speed of the particles. In fact, temperature is a measure of the average speed of the particles.
- The higher the temperature, the larger the volume of a gas at constant pressure.

- There is a considerable expansion of a gas with increases in temperature at constant pressure, e.g. the volume of a gas increases by about 27% when it is heated from 20°C to 100°C.

Skills

Drawing simple particle diagrams

Draw a diagram to show the arrangement and motion of particles of a gas. Use the model of Figure 2.3 to guide you.
- Draw the particles as circles well spread out and arranged randomly through the container.
- Label the circles 'particles move at random colliding with the walls of the container and each other'.

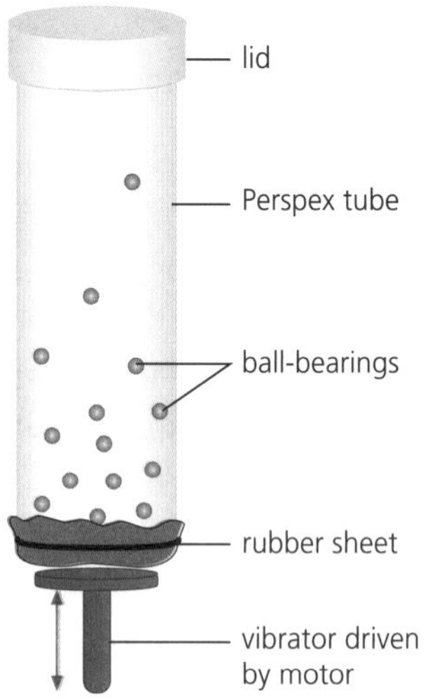

Brownian motion

In the apparatus in Figure 2.4, smoke particles reflect the light, which is seen in the microscope as tiny bright dots. They move around randomly, and they also appear and disappear as they move vertically. This movement is caused by the irregular collisions between the microscopic smoke particles and fast-moving, **invisible** air particles. This is clear evidence for the particle model of matter.

lid

Perspex tube

ball-bearings

rubber sheet

vibrator driven by motor

▲ **Figure 2.3 A model of the behaviour of the particles of a gas**

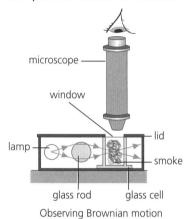

microscope

window

lid

lamp

smoke

glass rod   glass cell

Observing Brownian motion

▲ **Figure 2.4 Observing Brownian motion of smoke particles**

The particle model was first observed by Robert Brown, who observed in a microscope pollen particles suspended in water moving haphazardly due to bombardment by fast-moving, invisible water molecules.

Smoke/pollen particles are visible in a microscope and relatively massive.

Air/water particles are invisible and are fast-moving molecules.

Sample questions

1 Compare:
   a the separation of particles of a liquid and a gas   [2]
   b the nature of the motion of particles of a solid and a liquid   [2]

   c the forces between the particles of a solid and a gas   [3]

Student's answers

   a   it is greater   [0]
   b   they vibrate   [1]

   c   there are strong attractive forces between particles of a solid but no
       forces between the particles of a gas   [2]

Teacher's comments

a This is a vague statement that gives no information.
b The student's statement is still too vague as it is not stated which state of matter it refers to. One mark is scored as the particles of both states do vibrate.

c This statement is correct for a solid and most of the time for a gas.

Cambridge IGCSE Physics Study and Revision Guide Third Edition

Correct answers

a   The particles of a gas are much further apart. [2]
b   The particles of both states vibrate but some particles of a liquid also move randomly through the liquid. [2]

---

c   There are strong attractive forces between particles of a solid. There are no forces between the particles of a gas except when they collide and there are strong repulsive forces. [3]

---

2   A student looks in a microscope at a cell containing illuminated smoke particles. Explain:
a   what is seen [1]
b   the movement observed [1]
c   what causes this movement [2]

Student's answers

a   *Smoke particles* [0]
b   *Moving around* [0]
c   *The smoke molecules are bombarded by air.* [1]

Teacher's comments

a   It is reflected light, not smoke particles, that is seen.
b   Moving around is too vague.
c   The word 'molecule' is incorrect and the whole answer is incomplete.

Correct answers

a   Bright specks of light. [1]
b   Moving around haphazardly in *all* directions. [1]
c   The bright specks are light reflected off the smoke particles, which are bombarded by air particles. [2]

2.1.3   Gases and the absolute scale of temperature

Key objectives

By the end of this section, you should be able to:
●   describe and explain in terms of particles the changes of gas pressure with changes of temperature and volume

●   recall and use the equation to convert temperature between kelvin and degrees Celsius

●   recall and use the equation pV = constant, including a graphical representation of this relationship

Gas pressure

There is a force when fast-moving particles collide with the walls of the container they are in. Gas pressure is caused by the total force of collisions per unit area. The higher the temperature, the faster the particles move. If the volume is kept constant, the pressure increases because:

●   there are more frequent collisions with the container walls

●   the collisions are harder, so exert more force

Note: The gas particles do collide with each other but this is not relevant to the cause of gas pressure.

Gas pressure and volume at constant temperature

At a **constant temperature**, gas molecules move at a constant average speed, so the average force from each collision is the same. If the gas is compressed into a smaller volume, there are more frequent collisions on each unit of area of the surface. So, the total force per unit area increases and the pressure increases.

Similarly, if the gas expands to a greater volume at a constant temperature, the pressure decreases.

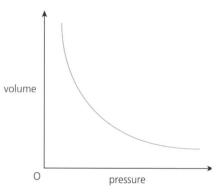

▲ Figure 2.5 Pressure and volume of a fixed mass of gas at constant temperature

You should know and be able to use the equation for a fixed mass of gas at constant temperature:

$$pV = \text{constant} \qquad \text{or} \qquad p_1V_1 = p_2V_2$$

The relationship between the varying pressure and volume of a fixed mass of gas at constant temperature is shown by Figure 2.5.

> **Skills**
>
> **Rearranging the equation to make any variable the subject**
>
> You need to be able to rearrange the equation $p_1V_1 = p_2V_2$ to make any of the variables the subject.
>
> To make p_1 the subject:
>
> $$p_1V_1 = p_2V_2$$
>
> $$p_1 = \frac{p_2V_2}{V_1}$$
>
> To make p_2 the subject:
>
> $$p_1V_1 = p_2V_2$$
>
> $$p_2V_2 = p_1V_1$$
>
> $$p_2 = \frac{p_1V_1}{V_2}$$
>
> To make V_1 the subject:
>
> $$p_1V_1 = p_2V_2$$
>
> $$V_1 = \frac{p_2V_2}{p_1}$$
>
> To make V_2 the subject:
>
> $$p_1V_1 = p_2V_2$$
>
> $$p_2V_2 = p_1V_1$$
>
> $$V_2 = \frac{p_1V_1}{p_2}$$

Celsius and Kelvin temperature scales

Absolute zero is the lowest possible temperature. On the Celsius scale absolute zero is −273°C. On the Kelvin scale it is 0 K. The two scales have units of the same size and are related by the equation:

$$\text{temperature in K} = \text{temperature in } °C + 273$$

> **Skills**
>
> **Converting between kelvin and degrees Celsius**
>
> You need to be able to convert between **kelvin** and **degrees Celsius**. To do this you need to rearrange the equation temperature in K = temperature in °C + 273 to make temperature in °C the subject.
>
> $$\text{temperature in } °C + 273 = \text{temperature in K}$$
>
> $$\text{temperature in } °C = \text{temperature in K} - 273$$

Sample questions

REVISED

3   A piston slowly compresses a gas from 540 cm³ to 30 cm³, so that the temperature remains constant. The initial pressure was 100 kPa. Find the final pressure.   [3]

Student's answer

Vol = 30 cm P = 100

100 × 30 = 3000 N   [0]

Correct answer

Note: If you set out your working logically, as shown below, you are much more likely to get the answer right.

$p_1 = 100 \, \text{kPa}$, $p_2 = \text{unknown}$

$V_1 = 540 \, \text{cm}^3$, $V_2 = 30 \, \text{cm}^3$

$p_1 V_1 = p_2 V_2$

$100 \times 540 = p_2 \times 30$

$p_2 = \dfrac{100 \times 540}{30} = \dfrac{54\,000}{30} = 1800 \, \text{kPa}$ [3]

Teacher's comments

The student has not approached the question at all systematically or used the right equation. Two numbers have been combined to give a value but this is meaningless and the unit is also incorrect for pressure.

4 A gas cylinder is heated in a fire. State what happens to the pressure of the gas and explain your answer in terms of the gas particles.   [4]

Student's answer

The pressure increases because the particles move around more, hitting each other and the walls.   [1]

Correct answer

The pressure increases because the particles move faster [2], hitting the walls more frequently and harder, thus increasing the total force on the walls per unit area.   [2]

Teacher's comments

The student's answer is vague, mentioning the particles colliding with each other, which is irrelevant.

5 Describe in terms of the particles of a gas the effect on the pressure of a gas of an increase of volume at constant temperature.   [4]

Student's answer

The pressure gets less because the particles are further apart.   [1]

Correct answer

The pressure decreases. The particles move with the same velocity but there are fewer collisions with the walls with the larger volume. So, there is less force on the walls per unit area.   [4]

Teacher's comments

The comment about pressure is correct. The rest of the statement is vague and does not explain anything.

6 The temperature in a cold store is −24°C. Calculate the temperature in kelvin.   [2]

Student's answer

temperature = 24 + 273 = 297 K   [1]

Teacher's comments

The student knew 273 should be added but used the wrong Celsius temperature.

Correct answer

temperature in K = temperature in °C + 273 = −24 + 273 = 249 K   [2]

Revision activities

1 Draw sketches to show models of the particles of solids, liquids and gases.
2 Draw a sketch of the path of smoke particles showing Brownian motion as would be seen in a microscope.

3 Make a revision poster for the equation showing the pressure and volume of a fixed mass gas which change at constant temperature. State the variables. Rearrange the equation four times so each variable is on its own on the left of the equals sign.

Exam-style questions

Answers available at: www.hoddereducation.co.uk/cambridgeextras
1 A gas particle strikes the wall of a container and bounces back. Explain:

   a in terms of momentum, how this causes a force on the walls of the container   [2]

   b how all the particles of the gas cause a pressure on the walls of the container   [2]
2 The volume of a gas increases at constant temperature. The particles of the gas are moving.
For parts **a** to **d**, choose **one** of *increases*, *decreases* or *stays the same* to describe the new state of each quantity.
   a pressure
   b kinetic energy of the particles
   c rate of collisions of particles per unit area of walls
   d total force per unit area   [4]
3 Which description compares the properties of the particles of solids to the properties of particles of gases?

	Solids	Gases
A	closest	move slowest
B	closest	move fastest
C	furthest apart	move fastest
D	furthest apart	move slowest

    [1]

4 An experiment is carried out on some gas contained in a cylinder by a piston, which can move.
In stage 1, the gas is heated with the piston fixed in position. State and explain whether the following will increase, decrease or stay the same during stage 1:
   a the speed of the gas particles   [2]
   b the number of collisions per second between particles and the walls   [2]
   c gas pressure   [2]

It is illegal to photocopy this page

5 A sample of gas stays at a fixed temperature while its volume increases.
Which describes the new state of the gas?

	Speed of the particles of the gas	Number of collisions per second between particles and walls
A	increases	stays the same
B	stays the same	decreases
C	stays the same	increases
D	decreases	stays the same

[1]

6 Describe an experiment using pollen particles to demonstrate Brownian motion. You should:
  a draw a labelled diagram of the apparatus   [2]
  b state what is seen   [2]
  c explain how what is seen illustrates Brownian motion   [2]

7 The volume of the container holding a fixed mass of gas is reduced from $800\,cm^3$ to $300\,cm^3$. The final pressure of the gas is $600\,kPa$ above the atmospheric pressure of $100\,kPa$.
Calculate the initial pressure of the gas compared with atmospheric pressure.   [4]

2.2 Thermal properties and temperature

REVISED ☐

Thermal energy flows from a hot body to a cold body.

Temperature measures the amount of thermal or internal energy in a body. In everyday terms, it measures how hot a body is.

2.2.1 Thermal expansion of solids, liquids and gases

Key objectives

By the end of this section, you should be able to:
● describe the consequences of the thermal expansion of solids, liquids and gases in a wide range of practical situations from everyday life
● describe the expansion of solids, liquids and gases as their temperatures rise

Relative amount of expansion of solids, liquids and gases
Solids when heated expand the least:

● Particles are close together and vibrate about fixed points in a regular array or lattice.
● As temperature increases, the particles vibrate further and faster. This pushes the fixed points a little further apart and the solid expands.

Liquids when heated expand more than solids but less than gases:

● Particles are slightly further apart than in solids but still close enough to keep a definite volume.
● The main motion of the particles is vibration. The particles also move randomly in all directions, not being fixed to each other.

- As temperature increases, the particles move faster and further apart, so there is a small expansion of a liquid.

Gases when heated expand the most:

- Particles are much further apart than in solids or liquids. They move much faster than in solids or liquids and move throughout the available space.
- The higher the temperature, the faster the speed of the particles.
- The higher the temperature, the larger the volume of a gas to keep the pressure constant.
- There is a considerable expansion of a gas with increases in temperature at constant pressure.

▼ Table 2.1 Uses and disadvantages of thermal expansion

	Uses	Disadvantages
Thermal expansion of solids	Shrink-fitting, curling of a bimetallic strip in a fire alarm	Gaps need to be left between lengths of railway line to allow for expansion in hot weather.
Thermal expansion of liquids	Liquid-in-glass thermometers	The water in a car's cooling system expands when the engine gets hot. A separate water tank is needed for the hot water to expand into.
Thermal expansion of gases	Internal combustion engines	Gas cylinders can explode if overheated.

In the fire alarm circuit in Figure 2.6, thermal energy from the fire causes the lower metal in the bimetallic strip to expand more than the upper metal. This causes the strip to curl up, which completes the circuit and the alarm bell rings.

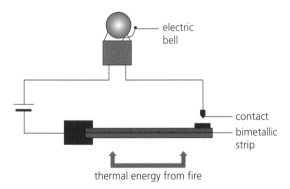

electric bell

contact

bimetallic strip

thermal energy from fire

▲ Figure 2.6 A fire alarm

Sample question

7 The lid is stuck on a glass jar. How could you use hot water to release it? Explain in terms of the particles how this works. [4]

Student's answer

> Put the glass jar in hot water and the lid will come off [1] because the molecules expand. [0]

Teacher's comments

The student did not specify where exactly the hot water should be used and gave a vague, incorrect explanation of the role of the molecules. It is quite acceptable to use the terms molecules or particles.

2.2.2 Specific heat capacity

Key objectives

By the end of this section, you should be able to:
- understand that a rise in an object's temperature increases its internal energy
  - describe temperature rise of an object in terms of an increase in average kinetic energy of its particles
- define specific heat capacity, and recall and use the equation for it
- describe experiments to measure the specific heat capacity of a solid and a liquid

When thermal energy flows into a body, its molecules move faster, increasing its internal energy, the kinetic energy of its particles and its temperature.

Specific heat capacity is a property of a material.

Specific heat capacity is defined as the thermal energy needed *per kilogram* (unit mass) to increase the temperature of a material by 1°C.

You should know and be able to use the following equation:

$$c = \frac{\Delta E}{m\Delta\theta}$$

You should know the following symbols: c = specific heat capacity, ΔE = energy change, m = mass, $\Delta\theta$ = temperature change.

Skills

Rearranging the equation for specific heat capacity
You need to be able to rearrange the specific heat capacity equation $c = \Delta E/m\Delta\theta$ to make any of the variables the subject.

Write down the equation when ΔE is the subject:

$$c = \frac{\Delta E}{m\Delta\theta}$$

$$\frac{\Delta E}{m\Delta\theta} = c$$

$$\Delta E = cm\Delta\theta$$

Write down the equation when m is the subject:

$$c = \frac{\Delta E}{m\Delta\theta}$$

$$cm\Delta\theta = \Delta E$$

$$m = \frac{\Delta E}{c\Delta\theta}$$

Write down the equation when $\Delta\theta$ is the subject:

$$c = \frac{\Delta E}{m\Delta\theta}$$

$$cm\Delta\theta = \Delta E$$

$$\Delta\theta = \frac{\Delta E}{cm}$$

Write down the units of specific heat capacity:

J/(kg °C)

Skills

Measuring the specific heat capacity of a solid

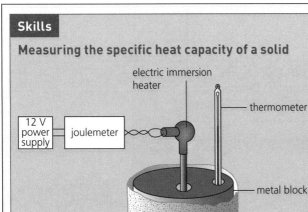

▲ **Figure 2.7 The insulation reduces thermal energy transfer to the surroundings**

Measure the mass of the metal block and the temperature before and after heating, and record the joulemeter reading of the energy supplied

(Figure 2.7). Use these results to calculate the specific heat capacity of the block.

Here is an example of working out from another experiment:

mass of metal block = 1.6 kg

temperature before heating = 21°C

temperature after heating = 66°C

increase of temperature = 45°C

joulemeter reading = 46 800 J

specific heat capacity

$$= \frac{\text{energy supplied by immersion heater}}{\text{mass} \times \text{temperature increase}}$$

$$= \frac{46\,800}{1.6 \times 45}$$

$$= 650 \text{ J/(kg °C) to 2 s.f.}$$

Skills

Measuring the specific heat capacity of a liquid

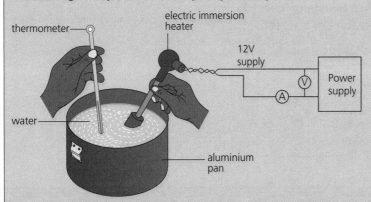

▲ **Figure 2.8 Experiment to find the specific heat capacity of a liquid**

Measure the mass of the water and the temperature before and after heating. Record the voltmeter (V) and ammeter (I) readings and the heating time in seconds (t). Use these results to calculate the specific heat capacity of the water.

energy received by water = VIt

specific heat capacity =

$$\frac{\text{energy received by water}}{\text{mass} \times (\text{temperature after heating} - \text{temperature before heating})}$$

Sample question

REVISED ☐

8 A heater of power 120 W heats a block of metal of mass 3.5 kg for 6 minutes. The specific heat capacity of the metal is 900 J/(kg°C) and its initial temperature 300 K. Calculate the final temperature of the metal block.   [4]

Student's answer

$$c = \frac{\Delta E}{m\Delta\theta}$$

energy supplied = ΔE = 120 × 6 = 720 J

$$\Delta\theta = \frac{m\Delta E}{mc} = \frac{3.5 \times 720}{900} = 2.8°C = 2.8\,K$$

final temperature = 273 + 2.8 = 276 K   [1]

Teacher's comments

The student started with the correct equation but made mistakes in applying it.

As watts are joules/second, the power should have been multiplied by the time in seconds.

The specific heat capacity equation was incorrectly rearranged for $\Delta\theta$.

The temperature rise was added to the Kelvin temperature of freezing water not to the initial temperature.

Correct answer

$$c = \frac{\Delta E}{m\Delta\theta}$$

energy supplied = ΔE = 120 × 6 × 60 = 43 000 J

$$\Delta\theta = \frac{m\Delta E}{mc} = \frac{43000}{3.5 \times 900} = 14°C = 14\,K$$

final temperature = 300 + 14 = 314 K   [4]

2.2.3 Melting, boiling and evaporation

Key objectives

By the end of this section, you should be able to:
- describe melting, boiling, condensation, solidification and evaporation and use these terms
- know the melting and boiling temperatures of water at standard atmospheric pressure
- know that evaporation causes cooling of a liquid
- describe the differences between boiling and evaporation
- describe how temperature, surface area and air movement affect evaporation, and explain the cooling effect of an evaporating liquid on an object

During **vaporisation** (**evaporation** and **boiling**) thermal energy is supplied to break the bonds between particles without a change of temperature. The boiling temperature of water at standard atmospheric pressure is 100°C.

Evaporation causes the particles in the remaining liquid to cool down because energy is needed to break the bonds between molecules, and the more energetic particles escape from the surface.

The rate of evaporation increases with:

- higher temperatures, as more particles at the surface are moving faster
- increased surface area, as more particles are at the surface
- a wind or air movement, as the gas particles are blown away so cannot re-enter the liquid

An object is cooled when in contact with an evaporating liquid because the liquid has cooled down.

Condensation occurs when gas or vapour particles return to the liquid state. Thermal energy is given out as the bonds between particles in the liquid re-form.

Melting, or fusion, takes place at a definite temperature called the melting point. Ice is solid water. The melting temperature of ice at standard atmospheric pressure is 0°C. Thermal energy must be provided to break the bonds between particles for them to leave the well-ordered structure of the solid.

Solidification, or freezing, occurs when particles of a liquid return to the solid state. This takes place at a definite temperature called the freezing point, which has the same value as the melting point. Thermal energy is given out as the bonds between particles of the solid re-form.

Differences between boiling and evaporation

Evaporation takes place from the surface of the liquid at all liquid temperatures.

Boiling occurs at a definite temperature called the boiling point. Bubbles of vapour form within the liquid and rise freely to the surface. Energy must be supplied continuously to maintain boiling.

Sample question

REVISED

9 A student is playing football on a cool, windy day, wearing a T-shirt and shorts. He feels comfortably warm because he is moving around vigorously. His kit then gets wet in a rain shower. Explain why he now feels cold. [2]

Student's answer

The wet T-shirt makes him feel cold. [0]

Correct answer

The water in his wet kit is evaporated by the wind. [1] The thermal energy needed for this evaporation is taken from the water in his T-shirt and shorts, as well as from his body, so he feels cold. [1]

Teacher's comments

The student's answer is far too vague and does not mention the cooling caused by evaporation.

Revision activity

Work in pairs to revise the equation connecting temperature in kelvin and °C and how to rearrange it. On your own, complete each of the following questions. Then swap answers and check each other's work.

1 Write down the equation connecting temperature in kelvin and °C with temperature in kelvin on its own on the left of the equals sign.

2 Write down the equation connecting temperature in kelvin and °C with temperature in °C on its own on the left of the equals sign.

Exam-style questions

Answers available at: www.hoddereducation.co.uk/cambridgeextras

8 An ice cube with a temperature of 0°C is placed in a glass of water (Figure 2.9) with a temperature of 20°C.
After a few minutes, some of the ice has melted. State and for (a) and (b) explain whether the following increase, decrease or stay the same:
   a the temperature of the remaining ice   [2]
   b the temperature of the water   [2]
   c the mass of water in the glass   [1]
   d the total mass of the ice and water   [1]

9 Write down three differences and two similarities between boiling and evaporation.   [5]

10 Write down the boiling temperature in kelvin and degrees Celsius of water at standard atmospheric pressure.   [2]

11 Two straight strips of metal alloys (invar and bronze) are bonded together at room temperature. Bronze expands appreciably when heated, but invar expands very little. Describe the shape of the strips when heated in an oven. Explain your answer.   [2]

12 An experiment is carried out to find the specific heat capacity of a metal. A 2 kg block of the metal is heated by a 200 W heater for 5 min, and the temperature of the block rises from 20°C to 51°C. Work out:
   a the energy supplied to the block by the heater   [2]
   b the specific heat capacity of the metal   [3]
   When used in an engine, a component made from this metal receives 35 kJ of thermal energy and its temperature rises from 30°C to 290°C.
   c Work out the mass of the component.   [2]

▲ Figure 2.9

2.3 Transfer of thermal energy

REVISED

Thermal energy is always transferred from a place of high temperature to a place of low temperature.

2.3.1 Conduction

Key objectives

By the end of this section, you should be able to:
- describe experiments to demonstrate the properties of good thermal conductors and bad thermal conductors
- describe thermal conduction in terms of lattice vibrations and in terms of the movement of free electrons
- describe why thermal conduction is bad in gases and most liquids, and know that many solids conduct thermal energy well

In **conduction**, thermal energy is transferred through a material without movement of the material.

> Metals are generally good conductors, but most other solids are poor conductors. Liquids are generally much worse thermal energy conductors than metals.
>
> Gases are all very poor conductors of thermal energy. For example, if you put your hands in very cold water, they will feel cold almost at once. If your hands are in air of the same temperature, they will cool down but at a much slower rate because air is a bad conductor.
>
> The atoms or molecules in a hot part of a solid vibrate faster and further than those in a cold part.
>
> In metals, thermal energy is transferred by fast-moving free electrons, which pass through the solid, causing atoms in colder parts to vibrate more.
>
> There is a secondary mechanism that is much slower. The vibrating atoms or molecules in the lattice cause their neighbours to vibrate more, thus passing on thermal energy. Non-metals do not have free electrons so can use only this mechanism, which is why non-metals are poor conductors. Some substances, e.g. semiconductors, have a limited number of free electrons so conduct thermal energy better than thermal insulators but less well than good thermal conductors.
>
> These mechanisms explain why liquids are generally poor conductors and gases even worse. There are no free electrons and the secondary mechanism works poorly. There is very little contact between vibrating atoms or molecules in liquids and almost none in gases.

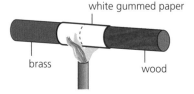

▲ **Figure 2.10 The paper over the brass does not burn because brass is a good conductor**

▲ **Figure 2.11 Water is a poor conductor of thermal energy**

Skills

Demonstrating the properties of good and bad thermal conductors
The practicals shown in Figure 2.10 and Figure 2.11 can be done to show the properties of good thermal conductors and poor thermal conductors.

2.3.2 Convection

Key objectives

By the end of this section, you should be able to:
● explain convection in liquids and gases and describe experiments to illustrate convection

In **convection**, thermal energy is transferred owing to movement of the liquid or gas itself. Convection cannot take place in a solid.

The liquid or gas expands on heating, so its density falls. The warmer and lighter liquid or gas rises to the cooler region, transferring thermal energy in the process.

Demonstrating convection

Figure 2.12 demonstrates a convection current. The heated air above the candle rises up the left-hand chimney and draws smoke from the lighted paper down the right-hand chimney and into the box.

Convection currents in water can be seen by dropping a potassium permanganate crystal into a beaker of water. The coloured traces indicate the flow of the convection currents.

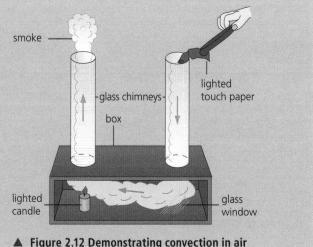

▲ **Figure 2.12 Demonstrating convection in air**

2.3.3 Radiation

Key objectives

By the end of this section, you should be able to:
- know that thermal radiation is infrared radiation, which does not require a medium
- describe the effect of surface colour and texture on the emission, absorption and reflection of infrared radiation

- describe experiments to distinguish between good and bad emitters and absorbers of infrared radiation
- describe how temperature and surface area affect rate of emission of radiation

- know the effect on the temperature of an object of the rates at which it receives and transfers energy away
- know how the Earth's temperature is affected by the balance between incoming radiation and radiation emitted from the Earth's surface

In **radiation**, thermal energy is transferred by infrared radiation, which is part of the electromagnetic spectrum (see Topic 3). All objects emit this radiation.

Surfaces that are good absorbers of thermal energy radiation are also good emitters. Surface colour and texture can affect emission, absorption and reflection of infrared radiation. Black surfaces are better emitters and absorbers than white surfaces. Dull or matt surfaces are better emitters and absorbers than polished surfaces.

There need not be any matter between the hot and cold bodies (no medium is required).

Most solids and liquids absorb infrared radiation, including water, which is transparent to light.

Core students do not have to describe the two experiments below, but if you understand them it will help you to answer questions on the Core syllabus.

The rate of radiation emitted increases with the surface area and surface temperature.

Skills

Good and bad emitters and absorbers of infrared radiation

Good and bad emitters

The copper sheet in Figure 2.13 has previously been heated strongly with a Bunsen burner. The hand next to the black surface feels much hotter than the hand next to the polished surface. This is because black surfaces emit thermal energy radiation more than polished surfaces.

Health and safety note: if you do this experiment, the copper plate needs to be very hot. If touched, it could cause a serious burn.

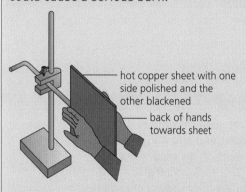

hot copper sheet with one side polished and the other blackened

back of hands towards sheet

▲ **Figure 2.13 Comparing emitters of radiation**

Good and bad absorbers

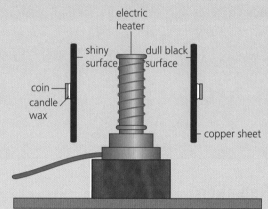

electric heater

shiny surface

dull black surface

coin
candle
wax

copper sheet

▲ **Figure 2.14 Comparing absorbers of radiation**

The heater is the same distance away from each copper sheet in Figure 2.14, so each receives the same amount of radiation. The dull black surface absorbs much more radiation than the shiny surface, so after a few minutes the wax on the black sheet melts and the coin falls off. The shiny surface reflects a lot of the radiation, so it stays cool and the wax does not melt.

Health and safety note: for clarity, the essential protective guard round the heater has not been shown. This safety feature is essential to protect users from a hot object at high electrical voltage.

If an object receives energy at a higher rate than it transfers it away, its temperature increases. If it receives energy at a lower rate than it transfers it away, its temperature decreases. If the two rates are the same, the object remains at constant temperature. The temperature of the Earth is determined by the balance between incoming radiation and radiation emitted.

2.3.4 Consequences of thermal energy transfer

Key objectives

By the end of this section, you should be able to:
- explain basic everyday applications of conduction, convection and radiation

- explain more complex applications of conduction, convection and radiation where more than one type of thermal energy transfer is significant

Saucepans and other solids through which thermal energy must travel are made of metals such as aluminium or copper, which are **good conductors**.

Blocks of expanded polystyrene are used for house insulation because they contain trapped air, which is a **bad conductor**.

A domestic radiator heats the air next to it which then rises and transfers thermal energy to the rest of the room. Despite its name, a radiator works mainly by **convection**.

Double glazing reduces the transfer of thermal energy by trapping a narrow layer of air between the window panes and **reducing convection**.

The Sun heats the Earth by infrared **radiation** through space.

Refrigerators have cooling pipes at the back. These have fins to give a larger surface area to increase loss of thermal energy by convection and radiation. The fins are also painted black to increase thermal energy loss by radiation because black surfaces are **good emitters**.

Many buildings in hot countries are painted white because white surfaces are **bad absorbers** of radiation from the Sun.

The colour of the surface influences radiation only. Black surfaces do not increase thermal energy transfer by conduction and convection.

> In more complex situations, more than one type of thermal energy transfer can be significant.
>
> A fire burning wood or coal warms a room by convection and radiation.
>
> The radiator of a car dissipates thermal energy by convection and radiation.

Sample questions

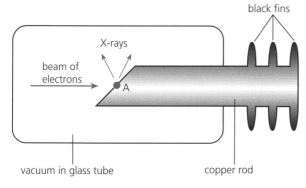

▲ **Figure 2.15**

10 Figure 2.15 shows an X-ray tube. Only a small proportion of energy from the electrons that strike point A goes into the X-rays that are emitted from that point. Most of the energy is transferred to thermal energy at point A; this energy is removed by the copper rod. Explain how conduction, convection and radiation play a role in the removal of this thermal energy. [6]

Student's answer

> The thermal energy goes down the rod and is conducted out by the fins [1] through convection and radiation. [1]

Teacher's comments

The student's answer shows an incorrect understanding of conduction.

Correct answer

Thermal energy is conducted along the rod to the fins [2] and is then emitted from the fins to the air by convection and radiation. [2] The black colour of the fins increases the rate of radiation. [2]

11 A metal spoon rests in a hot drink in a cup.
   a State the type of thermal energy transfer which makes the end of the spoon out of the drink become hot. [1]

   b Explain in terms of electrons and lattice vibrations why the end of a wooden spoon would not become so hot. [4]

Student's answer

   a The electrons are conducted along the spoon. [1]

   b Lattice vibrations occur in solids. The electrons in the metal are free to move and transfer energy to the end of the metal spoon. [2]

Teacher's comments

a Although it is an incorrect statement and electrons are not conducted, the mark is scored for the mention of conduction.

b The student has scored some marks for correct mentions of lattice vibration and free electrons. However, this was not applied to the situation and did not relate to how the ends of the spoons became hot.

Correct answer

   a conduction [1]

   b Lattice vibrations transfer thermal energy in all solids but this is a poor energy transfer mechanism. So the end of the wooden spoon receives little thermal energy and does not become hot. The free electrons in the metal move readily and transfer much more thermal energy so the end of the metal spoon becomes much hotter. [4]

12 Heat is applied to the bottom of a pan containing water. State and explain any types of thermal energy transfer which are significant in heating the water. [4]

Student's answer

Conduction and radiation. [1]

Correct answer

Thermal energy is conducted through the pan. This heats the water at the bottom, which transfers thermal energy to the rest of the water by convection. [4]

Teacher's comments

Conduction does occur which scores a mark. Radiation is not significant so is an incorrect answer. There is no explanation about where the types of energy transfer take place.

Revision activity

Write down how the change of density causes convection in a liquid or gas.

Revision activity

Make flash cards for the following questions to revise the mechanisms for conduction of thermal energy in solid metals and thermal radiation. Write the question on one side of each card and the answer on the other.

1 Explain how thermal energy is conducted by electrons in metals.
2 Explain the secondary mechanism of conduction used by non-metals.
3 Write down what is meant by thermal radiation.
4 State and explain whether thermal radiation can occur in space.

Exam-style questions

Answers available at: www.hoddereducation.co.uk/cambridgeextras

13 Explain in terms of particles why metals are much better thermal conductors than gases and most liquids. [3]

14 A child goes for a walk in winter in a cold country.
   a They open a metal gate which makes their hands cold. State and explain which type of thermal energy transfer cools their hands. [2]
   b They then wash their hands in a stream. State and explain which type of thermal energy transfer cools their hands. [2]
   c When they go home, they find a fire has been lit in the fireplace, and so hold out their hands near the fire. State and explain which type of thermal energy transfer warms their hands. [2]

15 Two space probes X and Y are identical in size and shape are in orbit around the Sun.
   a State the type of thermal energy transfer from the Sun to the probes. [2]
   b Probe X has a dull black surface and probe Y a shiny black surface. Which probe absorbs thermal energy from the Sun at a higher rate? [1]

   c Probe X is moved to an orbit further from the Sun. State and explain what happens to the steady temperature of probe X in its new orbit. [3]

16 In cold countries, houses are often heated by metal objects filled with hot water, often referred to as 'radiators'.
   State and explain the types of thermal energy transfer involved in the energy transfer from the hot water to the air in a house. [4]

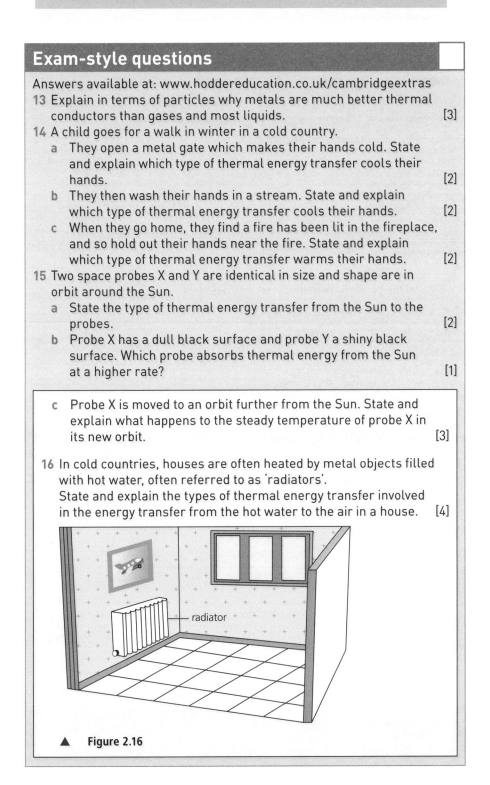

▲ **Figure 2.16**

3 Waves

Key terms

Term	Definition
Amplitude	The maximum displacement of a wave from the undisturbed position, or maximum change of value from zero
Converging lens	A lens which refracts parallel rays of light such that they converge to meet at a point
Crest of a wave	A wavefront where all the points have their highest displacement
Diverging lens	A lens which refracts parallel rays of light such that they diverge away from a point
Electromagnetic spectrum	Waves of the same nature with a wide range of wavelengths made up of oscillating electric and magnetic fields
Focal length	The distance between the optical centre and the principal focus of a lens
Frequency	The number of complete oscillations per second
Longitudinal wave	Direction of vibration of particles of the transmitting medium is parallel to the direction of travel of the wave
Principal focus (focal point)	Point on the principal axis to which light rays parallel to the principal axis converge, or appear to diverge from
Real image	An image which can be formed on a screen
Transverse wave	Direction of vibration or change of value is perpendicular to the direction of travel of the wave
Trough of wave	A wavefront where all the points have their lowest displacement
Virtual image	An image which cannot be formed on a screen
Wave speed	The distance moved by a point on a wave in 1 s
Wavefront	A line on which the particles or values of the wave are in phase
Wavelength	The distance between corresponding points in successive cycles of a wave
Analogue signal	A signal that can take any value within a range
Compression	Regions where particles of the transmitting medium are closer together
Digital signal	A signal that can only take one of two definite values: high (maximum value) or low (close to 0)
Rarefaction	Regions where particles of the transmitting medium material are further apart
Refractive index	The ratio of the speeds of a wave in two different regions

3.1 General properties of waves

Key objectives

By the end of this section, you should be able to:
- know that waves transfer energy without transferring matter
- describe what is meant by wave motion as illustrated by simple experiments
- understand the terms wavefront, wavelength, frequency, crest (peak), trough, amplitude and wave speed
- recall and use the wave equation: $v = f\lambda$

- distinguish between transverse and longitudinal waves and know different types of each wave
- describe how water waves can be used to illustrate reflection, refraction and diffraction

---

- describe how wavelength and gap size affect diffraction through a gap
- describe compressions and rarefactions
- describe how, with increasing wavelength, there is less diffraction at an edge

---

Waves transfer energy from one point to another without transferring matter. Some waves (e.g. water waves and sound waves) are transmitted by particles of a material vibrating about fixed points. They cannot travel through a vacuum.

Electromagnetic waves (e.g. light waves and X-rays) are a combination of travelling electric and magnetic fields. They *can* travel through a vacuum.

Types of wave motion

In **transverse waves**, the oscillation of the material or field is at right angles to the direction of travel of the wave. Figure 3.1 shows a transverse wave travelling in a horizontal rope. Each piece of rope oscillates vertically about a fixed point, but the pieces do not oscillate in time with each other.

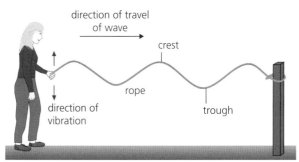

▲ **Figure 3.1 Transverse waves in a rope**

Transverse waves only oscillate vertically when the wave travels horizontally. Transverse waves oscillate horizontally when the wave travels vertically because the oscillation is always at right angles to the direction of travel.

In **longitudinal waves**, the oscillation of the material is parallel to the direction of travel of the wave.

Figure 3.2 represents a longitudinal wave travelling in a horizontal spring. Each coil of the spring oscillates horizontally about a fixed point, but the coils do not oscillate in time with each other.

The points marked 'C' are where the coils are most tightly packed (**compressions**) and 'R' marks the points where the coils are furthest apart (**rarefactions**).

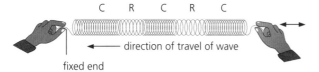

▲ **Figure 3.2 Longitudinal waves in a spring**

Longitudinal waves only oscillate horizontally when the wave travels horizontally. Longitudinal waves oscillate vertically when the wave travels vertically, because the oscillation is always parallel to the direction of travel.

Electromagnetic radiation, water waves and seismic S-waves (secondary) are transverse waves.

Sound waves and seismic P-waves (primary) are longitudinal waves.

Describing waves

The **wave speed** (v) is the distance moved by a point on the wave in 1 s.

The **frequency** (f) of a wave is the number of complete cycles per second and is measured in hertz (Hz).

The **wavelength** (λ) of a wave is the distance between two corresponding points (e.g. crests) in successive cycles.

The **amplitude** of a wave is the maximum displacement of the wave from the undisturbed position (marked a in Figure 3.3) or maximum change of value from zero.

Amplitude is not the height difference between the top of a crest and the bottom of a trough. Amplitude is the height difference between the top of a crest and the mean position, or between the bottom of a trough and the mean position.

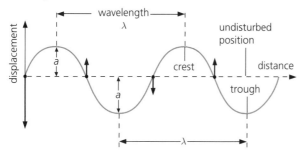

▲ **Figure 3.3 Displacement–distance graph for a wave at a particular instant**

Speed, frequency and wavelength are related by the equation:

$v = f\lambda$

Table 3.1 summarises some types of waves.

▼ **Table 3.1 Types of waves**

Type of wave	Longitudinal/transverse	Travel through a material
Wave on a rope	Transverse	Material needed
Wave on a spring	Either	Material needed
Water	Transverse	Material needed
Earthquake wave	Both	Material needed
Sound	Longitudinal	Material needed
Electromagnetic, e.g. light, X-rays, radio waves	Transverse	No material needed, but some electromagnetic waves can travel through certain materials

You need to be able to use the term **wavefront** as a line showing the position of a wave. A wavefront shows similar points of an extended travelling wave, such as a wave in water. The **crest** of a wave is a wavefront where all the points have their highest displacement. The **trough** of a wave is a wavefront where all the points have their lowest displacement.

Water waves

We can observe a wave travelling on the water surface of a ripple tank to illustrate how waves behave.

Key features of a ripple tank:

● A beam just touching the surface vibrates vertically to produce a wave.
● A light source shines through the water and shows the wave pattern on a screen above or below the ripple tank.

Reflection

In Figure 3.4, the wave produced by the vibrating beam is reflected from the flat metal barrier. The reflected wave is at the same angle to the reflecting surface as the incident wave. Speed, wavelength and frequency are unchanged by reflection.

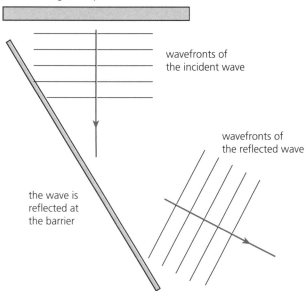

vibrating beam produces a wave

wavefronts of the incident wave

wavefronts of the reflected wave

the wave is reflected at the barrier

▲ **Figure 3.4 Reflection of a wave in a ripple tank**

Skills

Drawing wavefronts to show reflection at a plane surface

You must be able to describe how water waves can be used to show reflection at a plane surface (barrier).

● Draw a straight line across the page to represent the barrier.
● Draw four wavefronts separated by 10 mm at an angle of 35° from the barrier striking the middle section of the barrier. They show the incident wave.
● Add an arrow at right angles to the wavefronts to show the direction of travel.
● Your diagram should look like Figure 3.5.
● Where each incident wavefront strikes the barrier draw a reflected wavefront at 35° the other way from the barrier.
● Add an arrow at right angles to the reflected wavefronts to show the direction of travel.

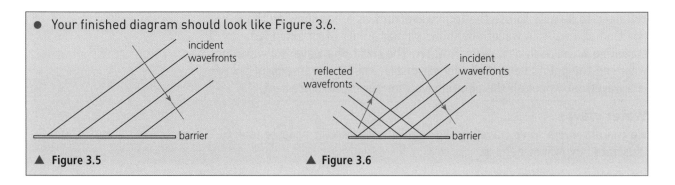

● Your finished diagram should look like Figure 3.6.

▲ **Figure 3.5**

▲ **Figure 3.6**

Refraction

Figure 3.7 shows a wave entering the shallow water above the glass where its speed is reduced. The frequency stays the same, so the wavelength is also reduced. The refracted wave changes direction.

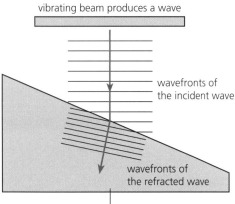

▲ **Figure 3.7 Refraction of a wave in a ripple tank**

Skills

Drawing wavefronts to show refraction due to change of speed

You must be able to describe how water waves can be used to show refraction.

In this exercise, you will draw a wave that is refracted as it travels into a region where its speed is changed due to a change in depth.

● Draw a line across the page at 30° to show the interface between the two regions where the wave has different speeds.
● Draw wavefronts horizontally at 30° to the interface separated by 10 mm and continue them down so three strike the interface.
● Add an arrow at right angles to the wavefronts to show the direction of travel.
● Your diagram should look like Figure 3.8.
● Draw the refracted wavefronts sloping up away from the barrier at 22° and joining onto the incident wavefronts where they strike the interface.
● Add an arrow at right angles to the refracted wavefronts to show the direction of travel.
● Your diagram should look like Figure 3.9.

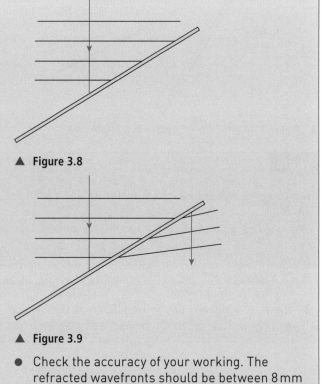

▲ **Figure 3.8**

▲ **Figure 3.9**

● Check the accuracy of your working. The refracted wavefronts should be between 8 mm and 9 mm apart measured at right angles to the wavefronts.

Diffraction through a narrow gap

Speed, wavelength and frequency are unchanged by diffraction, as shown in Figure 3.10.

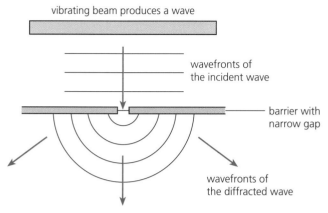

▲ **Figure 3.10 Diffraction of waves in a ripple tank by a gap that is narrower than the wavelength**

Notes for drawing wave diagrams

Points to note when drawing wave diagrams to show reflection, refraction and diffraction:

● Careful, accurate measuring and drawing are *essential* to produce good diagrams.

● The initial wavefronts must be parallel and have constant wavelength.

● Measure the wavelength of the incident waves.

● Reflected wavefronts must be parallel and have the same constant wavelength as the incident waves.

● Refracted wavefronts must be parallel and have a constant wavelength. Depending on the situation, the wavelength will be more or less than the incident wavelength.

● Diffracted wavefronts have straight and/or circular portions. The wavelengths between diffracted wavefronts must be carefully measured to be the same as the incident wavelength. The change of radius of the circular portions is the wavelength.

Diffraction at a gap or at an edge

Diffraction from a gap is shown in Figure 3.11.

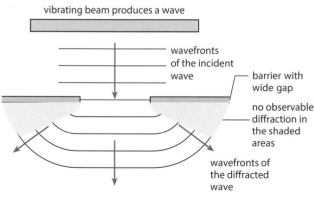

▲ **Figure 3.11 Diffraction of waves in a ripple tank by a gap**

Note: the centres of the part-circles are at the edges of the wide gap.

Diffraction at an edge is shown in Figure 3.12.

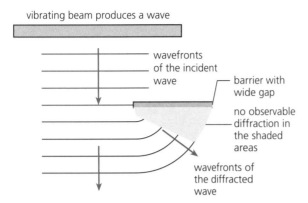

▲ **Figure 3.12 Diffraction of waves in a ripple tank by an edge**

Wavelength and gap size affect diffraction
Diffraction from a narrow gap is shown in Figure 3.13.

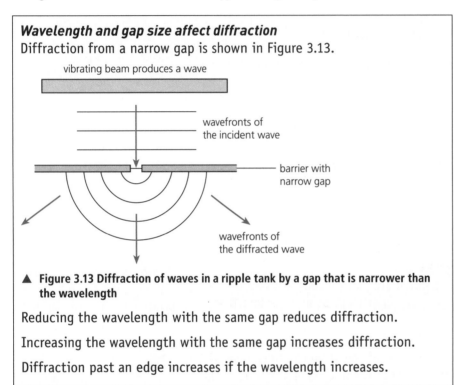

▲ **Figure 3.13 Diffraction of waves in a ripple tank by a gap that is narrower than the wavelength**

Reducing the wavelength with the same gap reduces diffraction.

Increasing the wavelength with the same gap increases diffraction.

Diffraction past an edge increases if the wavelength increases.

Skills

Describing diffraction due to an edge
Set up the ripple tank as shown in Figure 3.14.

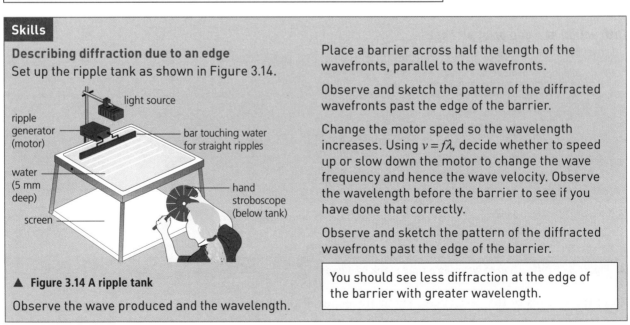

▲ **Figure 3.14 A ripple tank**

Observe the wave produced and the wavelength.

Place a barrier across half the length of the wavefronts, parallel to the wavefronts.

Observe and sketch the pattern of the diffracted wavefronts past the edge of the barrier.

Change the motor speed so the wavelength increases. Using $v = f\lambda$, decide whether to speed up or slow down the motor to change the wave frequency and hence the wave velocity. Observe the wavelength before the barrier to see if you have done that correctly.

Observe and sketch the pattern of the diffracted wavefronts past the edge of the barrier.

You should see less diffraction at the edge of the barrier with greater wavelength.

Sample questions

1 Sketch one and a half cycles of a transverse wave and mark on
  your sketch the amplitude and wavelength. [4]

Student's answer

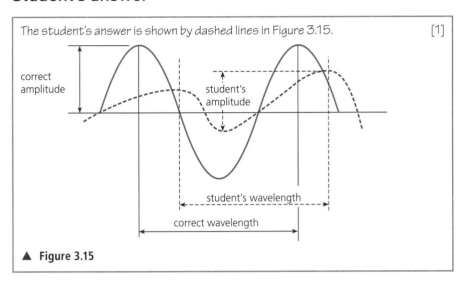

The student's answer is shown by dashed lines in Figure 3.15. [1]

correct amplitude

student's amplitude

student's wavelength

correct wavelength

▲ **Figure 3.15**

Teacher's comments

With this type of question, it is essential to work carefully and accurately
or few marks will be gained. Even for a sketch, the student's diagram is too
casual. For minimum acceptable accuracy, the student must indicate that
each half wavelength is the same length, and the distance from the axis to
each crest and trough is the same. The student's amplitude is incorrectly
measured from crest top to trough bottom. Given the irregular wave, the
student has correctly labelled one wavelength.

Correct answer

The correct answer is shown by the solid red line in Figure 3.15. [4]

2 A sensor detects that 1560 cycles of a wave pass in 30 s. Work out the
  frequency of the wave. [3]

Student's answer

frequency = 52 cycles in 1 s [2]

Teacher's comments

The student's answer is correct, but the unit of frequency is hertz (Hz).

Note that, although it is a fairly easy calculation, there is no working. If
the student had made a slight slip, no credit could have been given for
using the correct method.

Correct answer

$$\text{frequency} = \frac{\text{number of cycles}}{\text{time}} = \frac{1560}{30} = 52 \text{ Hz}$$

[3]

3 Find the frequency of a radio wave with a wavelength of 1500 m. [3]

> Please note, extended candidates are expected to know the speed of electromagnetic waves.

Student's answer

$v = f\lambda$ [1]

So

$f = \dfrac{v}{\lambda} = \dfrac{3 \times 10^8}{\lambda} = 200\,000\,\text{kHz}$ [1]

Correct answer

$v = f\lambda$ [1]

So

$f = \dfrac{v}{\lambda} = \dfrac{3 \times 10^8}{\lambda} = 200\,000\,\text{Hz} = 200\,\text{kHz}$ [2]

Teacher's comments

The student has done everything correctly, but the unit is wrong. It should be Hz not kHz. Perhaps the student assumed that, as radio frequencies are often expressed in kHz, this was the correct unit.

4 An earthquake wave is travelling vertically down into the Earth; the oscillations are also vertical. State, with a reason, whether the wave is longitudinal or transverse. [2]

Student's answer

The wave is transverse because it is vibrating up and down. [0]

Correct answer

The wave is longitudinal because the oscillations are parallel to the direction of travel. [2]

Teacher's comments

It is the direction of oscillation relative to the direction of travel that matters – the student does not mention this.

5 Complete the diagram to show the wave reflected at the barrier. [3]

▲ Figure 3.16

Student's answer

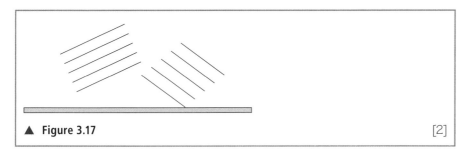

▲ **Figure 3.17** [2]

Correct answer

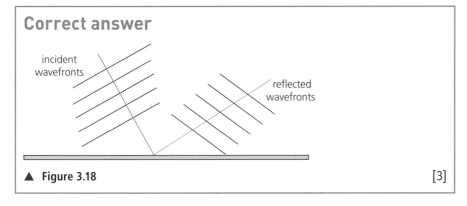

incident wavefronts

reflected wavefronts

▲ **Figure 3.18** [3]

Teacher's comments

The reflected wavefronts drawn by the student are all parallel and at the correct angle, which is good work. However, the wavelength between wavefronts is not constant. The wavelength must be the same as the incident wave.

6 Waves in a ripple tank have a wavelength of 16 mm and approach an interface parallel to the wavefronts. The waves slow down after the interface to 0.75 of their original speed.
  a Draw four wavefronts and the interface. [2]
  b Calculate the new wavelength after passing the interface. [2]
  c Add four wavefronts after passing the interface to your diagram drawn for part **a**. [2]

Student's answer

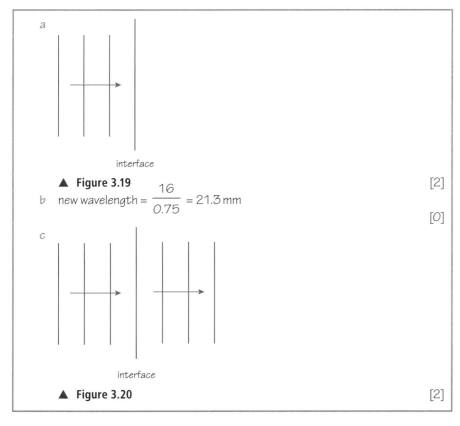

a

interface

▲ **Figure 3.19** [2]

b new wavelength $= \dfrac{16}{0.75} = 21.3\,\text{mm}$ [0]

c

interface

▲ **Figure 3.20** [2]

Teacher's comments

> a The student's answer is correct and carefully drawn.
> b The student has used inverse proportionality between wavelength and speed instead of direct proportionality.
> c Based on the incorrect wavelength in part b, the student has correctly drawn the new wavefronts.

Correct answers

a As student's answer. [2]

b As frequency is constant, wavelength is proportional to speed.
new wavelength = $16 \times 0.75 = 12$ mm [2]

c

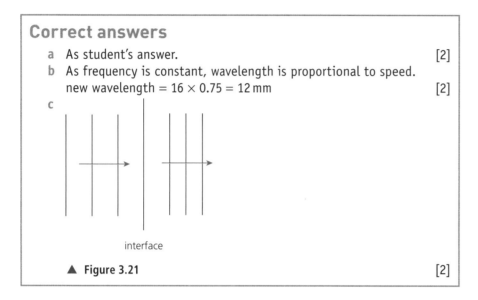

▲ Figure 3.21 [2]

7 Figure 3.22 shows wavefronts 12 mm apart approaching a barrier with a gap of 8 mm.
Draw carefully three wavefronts to show the pattern of the waves after passing through the barrier.

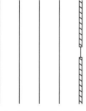

▲ Figure 3.22 [3]

Student's answer

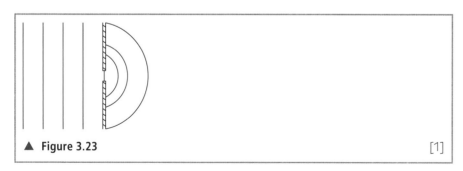

▲ Figure 3.23 [1]

Teacher's comments

The student has carefully drawn three semicircles which score 1 mark.

The semicircles should be centred on the middle of the gap. The wavelength should be constant and the same as the wavelength of the incident wavefronts.

Correct answer

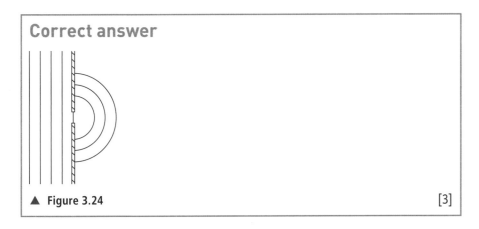

▲ Figure 3.24   [3]

Revision activity

Draw two careful sketches to show how wavefronts of different wavelength are diffracted differently as they pass the edge of a barrier.

Revision activity

Draw careful sketches to show the diffraction of water waves through a gap that is smaller than the wavelength. Then do the same for a gap that is much larger than the wavelength.

Exam-style questions

Answers available at: www.hoddereducation.co.uk/cambridgeextras

1 A woman swimming in the sea estimates that when she is in the trough between two crests of a wave, the crests are 1.5 m above her.
   a Work out the amplitude of the wave.   [1]
   b An observer counts that the swimmer moves up and down 12 times in 1 min. Work out the frequency of the wave.   [2]

2 A child throws a ball into a pond, hears the sound of the splash and observes water waves travelling towards him.
   a As the sound waves travel towards him, in which direction are the air particles oscillating?   [1]
   b As the water waves travel towards him, in which direction are the water particles oscillating?   [1]

3 A wave in water with a wavelength of 15 mm approaches a straight interface. The wavefronts are at an angle of 50° to the interface. After the interface, the speed of the wave increases. Draw three wavefronts before and three wavefronts after the interface to show any change of direction of the wave.   [4]

4 Figure 3.25 shows wavefronts of a wave approaching a barrier with a gap much larger than the wavelength.

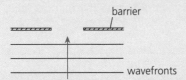

barrier

wavefronts

▲ Figure 3.25

   a Complete the diagram drawing three wavefronts which show the pattern of the waves after passing through the gap.   [3]
   b State the name of the process the wave undergoes.   [1]
   c Describe how the pattern of the wave after the gap would appear if the size of the gap was doubled.   [2]

3.2 Light

Light moves as waves of very small wavelength, but it is often convenient to use light rays to work out and explain the behaviour of light.

A light ray is the direction in which light is travelling and is shown as a line in a diagram.

An object is what is originally observed.

An optical image is a likeness of the object, which may not be an exact copy.

A real image is formed where the rays cross and can be shown on a screen.

A virtual image is observed where rays appear to come from and cannot be formed on a screen.

3.2.1 Reflection of light

Key objectives

By the end of this section, you should be able to:
- know and use the terms normal, angle of incidence, angle of reflection
- know what is meant by real and virtual images
- describe the formation of an optical image by a plane mirror

- state and use the relationship 'angle of reflection equals angle of incidence'

- perform simple constructions, measurements and calculations for plane mirrors

When light rays strike a mirror or similar surface, they return at the same angle from the normal as the incident ray. This is called **reflection**.

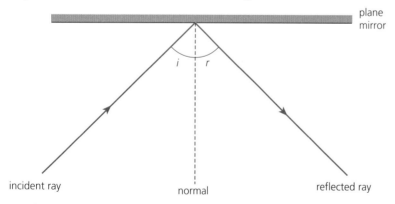

▲ **Figure 3.26 Reflection of light by a plane mirror**

i = angle of incidence = angle between **normal** and incident ray

r = angle of reflection = angle between **normal** and reflected ray

angle of incidence = angle of reflection or $i = r$

Real and virtual images

A **real image** is formed where light rays actually converge. It can be formed on a screen.

A **virtual image** can be seen as a point from which light rays diverge. It cannot be formed on a screen.

Formation of a virtual optical image by a plane mirror

Figure 3.27 shows the formation of a virtual image by a plane mirror.

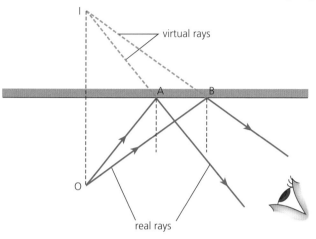

▲ **Figure 3.27 Construction to find the image in a plane mirror**

The image of the object O is not formed on a screen. It is at point I where the rays appear to come from. The properties of an image in a plane mirror are:

● It is the same size as the object.

● The line joining the object and the image is perpendicular to the mirror.

● It is the same distance behind the mirror as the object is in front of the mirror.

● It is laterally inverted.

● It is virtual.

Simple construction for reflection by a plane mirror

You must be able to draw simple constructions. Hints for drawing a construction to show the position of the image of a point object in a plane mirror:

● Carefully measure the distance of the object from the mirror.

● Mark the image the same distance behind the mirror as the object is in front of the mirror. The object and image should be on a line at right angles to the mirror line (OI in Figure 3.27).

● Draw two lines from the image towards the eye; draw dotted lines behind the mirror where they represent virtual rays.

● Join up the two lines from the object to where the previous two lines cut the mirror line (A and B in Figure 3.27).

● Mark arrows on the real rays and label the diagram as necessary.

Sample question

REVISED ☐

8 Draw a diagram to show the path of a ray striking a plane mirror with an angle of incidence of 35°. Mark and label the incident ray, normal, reflected ray and angles of incidence and reflection. [4]

Student's answer

The student's answer is shown by green dashed lines in Figure 3.28. [2]

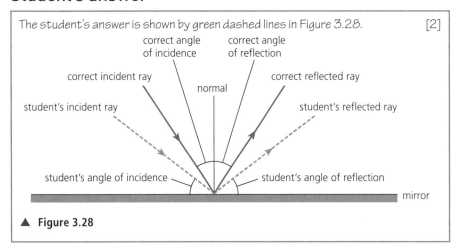

▲ **Figure 3.28**

The student has measured the angle of incidence away from the mirror line, not away from the normal. The normal is correct, as is the reflected ray for the incident ray drawn. The angle of reflection is also incorrectly measured away from the mirror line.

Correct answer

The correct answer is shown by the red solid lines in Figure 3.28. [4]

3.2.2 Refraction of light

Key objectives

By the end of this section, you should be able to:

- know and use the term angle of refraction
- describe an experiment to show the refraction of light by transparent blocks of different shapes, showing the passage of light through transparent material

- know that refractive index n = the ratio of the speeds of a wave in two different regions
- know and use the equation $n = \dfrac{\sin i}{\sin r}$

- state the meaning of critical angle

- know and use the equation $n = \dfrac{1}{\sin c}$

- describe internal reflection and total internal reflection with examples

- describe the use of optical fibres in telecommunication.

When a ray is travelling at an angle to a surface and enters a material where it travels slower, it changes direction *towards* the normal (Figure 3.29). This is called **refraction**.

When a ray leaves this material, it is refracted *away from* the normal.

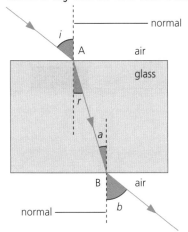

▲ **Figure 3.29 Refraction of a ray through a glass block**

Cambridge IGCSE Physics Study and Revision Guide Third Edition

For refraction at point A, where the ray enters the glass:

- i = angle of incidence = angle between **normal** and incident ray
- r = angle of refraction = angle between **normal** and refracted ray
- The ray is refracted towards the normal, so the angle of refraction is less than the angle of incidence.

For refraction at point B, where the ray leaves the glass:

- a = angle of incidence = angle between **normal** and incident ray
- b = angle of refraction = angle between **normal** and refracted ray
- Passing from glass to air, the ray is refracted away from the normal, so the angle of refraction is greater than the angle of incidence. When the block is parallel sided, the ray leaving is parallel to the ray entering.

Skills

Demonstrating refraction of light
You will need to be able to describe an experiment to demonstrate the refraction of light.

You should do this experiment in a darkened room. Direct a narrow beam of light at an angle to the side of a glass block placed on a large piece of paper. Mark on the paper the paths of the beams entering and leaving the block. By joining up the lines after removing the block, you can draw the path of the light as it travelled through the block. Figure 3.29 shows the paths of the rays in this experiment.

Measure the angles of incidence and refraction. Be careful to measure them from the normal not the interface.

This can be done with blocks of different shapes.

Refractive index
The amount of refraction is determined by the **refractive index** n – the ratio of the speed of light in air to the speed of light in the material.

For example, to find the refractive index of glass,

$$n = \frac{\text{speed of light in air}}{\text{speed of light in glass}}$$

The refractive index is related to the angles of incidence and refraction by this equation:

$$n = \frac{\sin i}{\sin v}$$

Internal reflection and critical angle
Figure 3.30 shows a ray inside a tank of water passing out into the air; some light is reflected internally and some is refracted away from the normal.

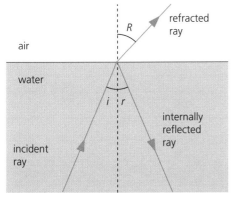

▲ **Figure 3.30 Rays at a water–air boundary**

i = angle of incidence = angle between **normal** and incident ray

r = angle of internal reflection = angle between **normal** and internally reflected ray

R = angle of refraction = angle between **normal** and refracted ray

The law of reflection still applies, so $i = r$.

The greater the angle of incidence, the more energy goes into the internally reflected ray, which becomes brighter. The greatest angle of incidence when refraction can still occur is called the **critical angle** (c). In Figure 3.31, the angle of refraction is 90° and the refracted ray travels along the surface.

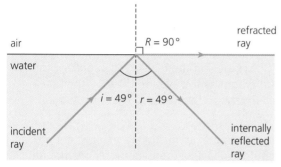

▲ **Figure 3.31 Angle of incidence is the same as the critical angle**

In this case, $\sin R = \sin 90 = 1$ and the refractive index is $1/n$ because the light is passing from water into light. The refraction equation becomes:

$$\frac{1}{n} = \frac{\sin c}{1} \text{ or } n = \frac{1}{\sin c}$$

If the angle of incidence is greater than the critical angle, there is no refracted ray and all of the energy is in the bright internally reflected ray. This is called **total internal reflection** (Figure 3.32).

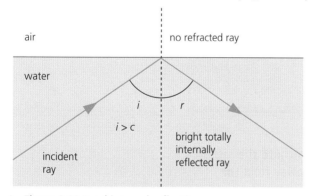

▲ **Figure 3.32 Total internal reflection**

Everyday uses of total internal reflection are binoculars and periscopes.

Skills

Internal reflection and total internal reflection in a semicircular glass block

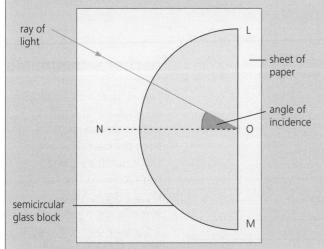

ray of light

L

sheet of paper

angle of incidence

N - - - - - - - - - - - - - - - - - O

semicircular glass block

M

▲ **Figure 3.33 Demonstrating total internal reflection**

- Direct the ray of light at an angle of incidence of about 20°.

  Observe the internally reflected ray emerging from the curved surface below the normal. Without measuring, notice that the angle of reflection is about equal to the angle of incidence.

Observe the refracted ray emerging from the flat surface at O. Without measuring, notice that the angle of refraction is greater than the angle of incidence.

- Direct the ray of light at an angle of incidence of about 35°.

  Observe the internally reflected ray emerging from the curved surface below the normal. The angle of reflection will still be about equal to the angle of incidence.

  Observe the refracted ray emerging from the flat surface at O. Without measuring, notice that the angle of refraction is now much greater than the angle of incidence.

- Slowly increase the angle of incidence from 35°.

  Observe the internally reflected ray emerging from the curved surface below the normal. The angle of reflection remains about equal to the angle of incidence. If you observe carefully, the reflected ray becomes a little brighter.

  The refracted ray emerges closer and closer to the flat surface at O. At some point there is no refracted ray. Total internal reflection has occurred. The angle of incidence at this point is the critical angle.

Optical fibres

Optical fibres are used in telecommunication with visible light or infrared, e.g. for cable television or high-speed broadband. Glass fibres are used because they are transparent to light and some infrared.

Each time the light strikes the wall of the optical fibre, the angle of incidence is greater than the critical angle and so total internal reflection occurs. There is very little loss of energy. The light can be considered trapped in the optical fibre and can travel long distances, even if the fibre is bent, in order to carry information or illuminate and view inaccessible places.

▲ **Figure 3.34 Light travels through optical fibre by total internal reflection**

Sample questions

9 Copy and complete Figure 3.35 to show the path of the ray through the glass prism as it is refracted twice. Show *both* normals. [4]

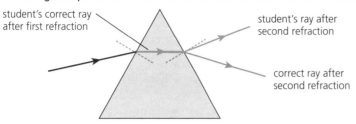

student's correct ray after first refraction

student's ray after second refraction

correct ray after second refraction

▲ **Figure 3.35 The black line shows the question**

Student's answer

> The student's answer is shown by the orange lines in Figure 3.35. [3]

Correct answer

The correct rays are shown by the green line in Figure 3.35. [4]

Teacher's comments

The student has correctly drawn the refracted ray within the prism and both normals. However, the student's second refraction, as the ray leaves the prism, is towards the normal. When a ray moves into a region where it travels faster, it is refracted away from the normal.

10 Light travels at 3×10^8 m/s in air and at 2.25×10^8 m/s in water. Calculate:
  a the refractive index, n, of water [2]
  b the angle of refraction for a ray approaching water with an angle of incidence of 55°. [2]

Student's answers

> a $\quad n = \dfrac{3 \times 10^8}{2.25 \times 10^8} = 1.33$ [2]
>
> b $\quad r = 34°$ [0]

Teacher's comments

> a Correct answer with working.
> b The answer is only slightly inaccurate, but there is no working, which means the examiner has no way of knowing whether the student has made a small mistake or was completely wrong and was simply close to the correct answer by chance. Examiners can only give credit for what they see.

Correct answers

> a $\quad n = \dfrac{3 \times 10^8}{2.25 \times 10^8} = 1.33$ [2]
>
> b $\quad \sin r = \dfrac{\sin i}{n} = \dfrac{\sin 55}{1.33} = \dfrac{0.8192}{1.33} = 0.616$
>
> $\quad r = 38°$ to 2 s.f. [2]

11 A ray of light is in water of refractive index 1.33. The ray approaches the interface with air at an angle of incidence of 52°. Carry out a suitable calculation and state what happens to the ray after striking the interface. [4]

Student's answer

$$n = \frac{\sin i}{\sin r}$$

$$\sin r = \frac{\sin 52°}{1.33} = 0.592$$

$$r = 36°$$ [1]

Teacher's comments

The student failed to make a final statement.

The student used the wrong value for refractive index.

Students should know and be able to use the equation $n = \frac{1}{\sin c}$.

Correct answers

For light passing from water to air, the correct value is $n = \frac{1}{1.33}$.

$$n = \frac{\sin i}{\sin r}$$

$$\frac{1}{1.33} = \frac{\sin i}{\sin r}$$

$$\sin r = 1.33 \times \sin 52° = 1.05$$

It is impossible to have a sin value greater than 1, which indicates that the critical angle must have been exceeded and total internal reflection occurs. This would be a completely correct answer. [4]

If a student happens to recognise that this is possibly the case, there is an alternative approach.

Alternative approach:

The angle of incidence is close to the critical angle, so check that first.

$$\sin c = \frac{1}{1.33}$$

$$c = 48.75°$$ [3]

The angle of incidence is confirmed as greater than the critical angle, so total internal reflection occurs and the ray is reflected back in the water with an angle of reflection of 52°. [1]

3.2.3 Thin lenses

Converging lenses

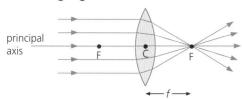

▲ Figure 3.36 Action of a converging lens on a parallel beam of light

All rays of light parallel to the principal axis are refracted by the **converging lens** to pass through the **principal focus**, F (Figure 3.36). The distance between F and the optical centre, C, is called the **focal length**, f.

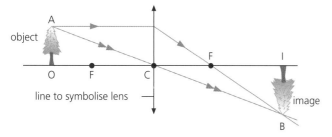

▲ Figure 3.37 Ray diagram for a converging lens

If the object is placed more than one focal length behind the lens, the image will always be real and inverted (Figure 3.37). Depending on where the object is placed, the image may be enlarged, the same size as the object or diminished.

Skills

Drawing ray diagrams

You must be able to draw ray diagrams to illustrate the formation of an image by a converging lens.

A lens has a focal length of 4.0 cm and an object of height 2.0 cm is placed 8.0 cm to the left of the lens.

Note: You must work **very carefully**. Small errors in drawing can lead to large errors in the result. Use a sharp pencil, draw points to within 0.5 mm of the correct position and draw lines **exactly** through points. Follow these steps to determine the position, size and nature of the image:

1 Preliminary – draw the principal axis, the object, a vertical line for the lens, mark the principal focus 'F' on both sides of the lens and mark the point where the principal axis crosses the lens as 'C'.

2 Ray 1 – draw a ray from the top of the object to the line of the lens, parallel to the principal axis, and continue this ray to pass through F and a few centimetres beyond. This is because all rays parallel to the principal axis pass through the principal focus.

3 Ray 2 – draw this ray from the top of the object through C to pass straight on until it cuts ray 1.

This is because the centre of the lens acts as a thin pane of glass, so rays pass through the centre undeviated.

4   The rays should converge at a point 8.0 cm to the right of the lens, 2.0 cm below the principal axis. Expect a small error of a few mm but no more. The image is real, inverted and the same size as the object.

5   The diagram should look like Figure 3.37.

Skills

Forming a real image with a converging lens

In a darkened room, set up on a bench a converging lens of known focal length, an illuminated object and a small white screen in a straight line as in Figure 3.38.

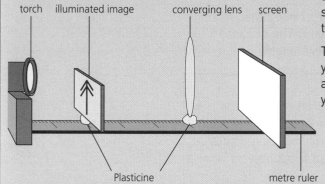

▲ **Figure 3.38 Forming an image with a lens**

Place the object about three focal lengths away from the lens and write down this distance. Place the screen on the other side of the lens about two focal lengths away from the lens. Observe the image as you slowly move the screen towards the lens. At the position of sharpest focus, write down the distance from the image to the lens.

To check your result, draw a ray diagram using your results as in Skills: Drawing ray diagrams and see if you get the same image distance as in your practical.

If the object is placed closer to a converging lens than the principal focus, the rays leaving the lens do not converge to form a real image. A virtual image is formed where diverging rays meet when extrapolated backwards. These rays cannot form a visible projection on a screen.

Magnifying glass

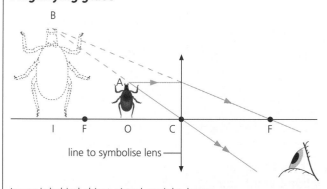

Image is behind object, virtual, upright, larger

▲ **Figure 3.39 Ray diagram for a converging lens used as a magnifying glass**

Figure 3.39 shows how a converging lens can be used as a magnifying glass. The object is placed less than one focal length behind the lens. No real image is formed, but the eye sees the rays diverging from the magnified virtual image. O is the object and I the image.

The nature of an image

The characteristics of an image are described using the following terms for an image when it is compared with the object: enlarged/the same size/diminished, upright/inverted and real/virtual.

Diverging lenses

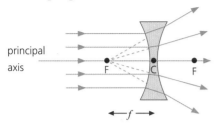

principal
axis

▲ **Figure 3.40 Action of a diverging lens on a parallel beam of light**

All rays of light parallel to the principal axis are refracted by the **diverging lens** to pass away from the principal focus, F. The distance between F and the optical centre, C, is called the focal length, f.

Lenses to correction vision

The human eye contains a converging lens which forms an image on the light-sensitive cells in the retina at the back of the eye. The muscles in the eye change the shape of the lens, which changes the focal length for objects at different distances. A healthy eye can focus objects from about 25 cm (Figure 3.41) to infinity (Figure 3.42).

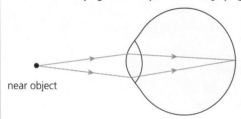

near object

▲ **Figure 3.41 Focusing an object at 25 cm, the eye lens is 'thicker' and has shorter focal length**

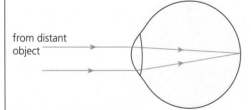

from distant
object

▲ **Figure 3.42 Focusing an object at infinity, the eye lens is 'thinner' and has longer focal length**

Short-sighted people have eye lenses that refract light from distant objects too much. They can see close objects clearly but distant objects are blurred. This is corrected by diverging lenses in spectacles or contact lenses (Figure 3.43).

Cambridge IGCSE Physics Study and Revision Guide Third Edition

a

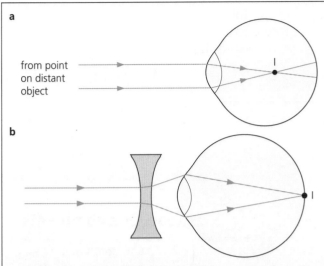

from point
on distant
object

b

▲ **Figure 3.43 Short-sightedness and its correction by a diverging lens**

Long-sighted people have eye lenses that do not refract light from near objects enough. They can see distant objects clearly but close objects are blurred. This is corrected by converging lenses in spectacles or contact lenses (Figure 3.44).

a

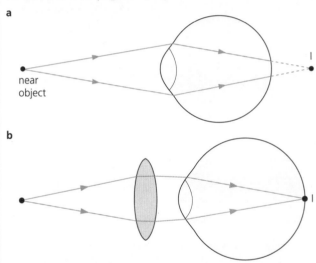

near
object

b

▲ **Figure 3.44 Long-sightedness and its correction by a converging lens**

Sample question

REVISED

12 a Figure 3.45 shows parallel rays from a distant object approaching the eye of a short-sighted person.

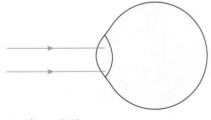

▲ **Figure 3.45**

Draw the rays continuing to show their paths to the back of the eyeball. [1]

b The person wears correcting lenses so distant objects can be seen in sharp focus.

In Figure 3.46 draw the correcting lens and the paths of the rays through the correcting lens and the eye lens to the back of the eyeball. [2]

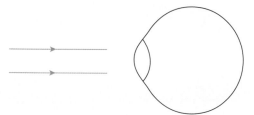

▲ Figure 3.46

Student's answer

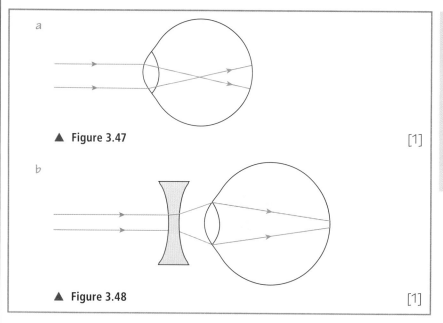

a

▲ Figure 3.47 [1]

b

▲ Figure 3.48 [1]

Teacher's comments

a The student's answer is correct.
b The student might have intended the correct answer but has not been careful when drawing. The rays should converge **exactly** at the back of the eyeball.

Correct answers

a The student's answer is correct. [1]

b

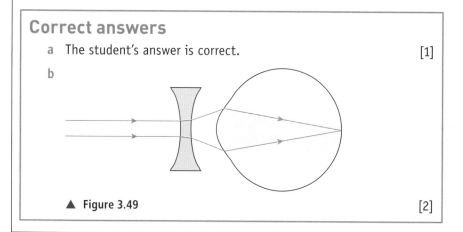

▲ Figure 3.49 [2]

3.2.4 Dispersion of light

Key objectives

By the end of this section, you should be able to:
- describe the dispersion of white light by a glass prism
- know the traditional seven colours of the visible spectrum in order of frequency and wavelength

- know that visible light of a single frequency is described as monochromatic

White light is made up of seven colours. In order of increasing wavelength these are violet, indigo, blue, green, yellow, orange and red. In order of increasing frequency, the sequence is reversed.

Each colour is refracted by a different amount in glass. If a beam of white light falls on a glass prism, it is dispersed into a **spectrum** of the seven colours (Figure 3.50).

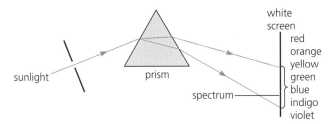

▲ **Figure 3.50 Forming a spectrum with a prism**

Monochromatic light
Light of a single frequency or wavelength is described as monochromatic, which means it is of a single colour.

Sample question

REVISED

13 Figure 3.51 shows a ray of green light passing through a glass prism from right to left.
A ray of orange light enters the prism on the same path as the original ray of green light.
On the diagram draw with a dashed line the path of this ray through and out of the prism. [3]

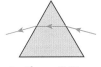

▲ **Figure 3.51**

Student's answer

▲ **Figure 3.52** [1]

Teacher's comments

The student has drawn the correct path within the prism. However, on leaving the prism the ray should be refracted away from the normal not towards the normal.

Correct answer

▲ Figure 3.53 [3]

Revision activity

Write down the seven colours in order of their refraction in a glass prism. You may find it helpful to use the capital letters of this mnemonic: Richard Of York Gave Battle In Vain (or the name 'Roy G Biv') to stand for the first letters of the colours. To remember which end of the visible spectrum is refracted most remember 'Blue bends best'.

Revision activity

Make flash cards to revise and rearrange the equation for refraction relating n, i and r with n on the left of the equals sign. Include what the symbols in the equation represent. Rearrange the equation with $\sin i$ on its own on the left of the equals sign. Then rearrange the equation with $\sin r$ on its own on the left.

Exam-style questions

Answers available at: www.hoddereducation.co.uk/cambridgeextras

5 Figure 3.54 shows a light ray entering a glass block of refractive index 1.5. The angle of incidence is changed so that angle A is now 60°.

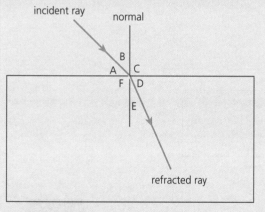

▲ Figure 3.54

    a Which of angles A–F is the angle of incidence? [1]
    b Which of angles A–F is the angle of refraction? [1]
    c Work out the values of angles B–F
      Do not measure from the diagram. [2]
6 a Describe the action of optical fibres in a medical application. [3]
    b Describe the action of optical fibres in a telecommunications application. [3]
7 An object of height 1.5 cm is placed 4.5 cm from a thin converging lens of focal length 3 cm. Draw a ray diagram to find:
    a the size and position of the image [4]
    b the nature of the image [3]

8 An object of height 1.5 cm is placed 2 cm from a thin converging lens of focal length 3 cm. Draw a ray diagram to show two parallel rays from the object passing through the lens.
    a What can you say about the two rays as they leave the lens? [1]
    b With dashed lines, draw your rays back behind the object. [2]
    c Find:
      i the size and position of the image [1]
      ii the nature of the image [1]

9 A surveyor's telescope used to observe a measuring pole with a scale produces an upside-down image compared with the object.
   a Suggest and explain a helpful way to have the numbers of the scale on the measuring pole printed. [1]
   b The telescope's image can be seen only in the eyepiece and cannot be projected on a screen.
   Choose three items from the following list that describe the image:

enlarged	inverted
same size	real
diminished	virtual [3]
upright	

10 a Draw a ray diagram to show the action of a diverging lens. [2]
   b Mark and label the principal axis and principal focus. [2]

11 Describe how a lens is used to correct long-sightedness. You should include a statement of the type of lens used and a ray diagram showing the action of the lens and the formation of a sharp image at the back of the eye. [4]

12 a Calculate the frequency of green light of wavelength 5.5×10^{-7} m. [3]

   b Use your knowledge of the electromagnetic spectrum to deduce and choose from the following list a possible wavelength for red light:
   5.5×10^{-10} m
   4.1×10^{-7} m
   7.1×10^{-7} m
   5.5×10^{-4} m [1]

3.3 Electromagnetic spectrum

REVISED

Key objectives

By the end of this section, you should be able to:
- know the main features of the electromagnetic spectrum

  - know that the speed of all electromagnetic waves in a vacuum is 3.0×10^8 m/s and is approximately the same in air

- describe typical uses of different regions of the electromagnetic spectrum
- describe the harmful effects on people of excessive exposure to electromagnetic radiation

- know that communication with artificial satellites is mainly by microwaves

  - know the important systems of communication using electromagnetic radiation
  - know the difference between a digital and an analogue signal and that sound can be transmitted as either
  - explain the benefits of digital signalling

Properties of electromagnetic waves
All types of waves that make up the **electromagnetic spectrum** have these properties in common:

- They can travel through a vacuum at the same high speed, which is much faster than other types of waves that travel through a material.

- They show the normal wave properties of reflection, refraction and diffraction.

- They are transverse waves.
- They travel owing to moving electric and magnetic fields.

> - The speed of electromagnetic waves in a vacuum is 3×10^8 m/s. The speed is approximately the same in air.

The wave equation $v = f\lambda$ applies, so the lower the wavelength, the higher the frequency.

The Sun and other stars give off a wide range of types of electromagnetic waves, which travel through space to Earth. Much of this radiation is stopped by Earth's atmosphere and can be detected only by satellites in orbit outside the atmosphere.

Types of electromagnetic waves

You need to know the types of electromagnetic waves, in order of decreasing wavelength and increasing frequency, as listed below.

Radio waves are used for:

- radio and television transmission
- radio astronomy detecting signals from stars and galaxies
- radio frequency identification (RFID) systems which track objects fitted with a small radio transmitter

> - Bluetooth signals which have limitations because they are weakened when passing through walls

Microwaves are used for telecommunication, radar and microwave ovens. They are used for satellite and mobile phone (cell phone) telecommunication because microwaves can pass through some walls and only require a short aerial. Microwaves can damage living cells and, as they travel through matter, cause internal burns. Parts of the ears and eyes, in particular, are easily damaged by microwaves, so great care is needed to ensure that the doors of microwave ovens are always closed when in use and that excessive mobile phone use is avoided. Personnel servicing military aircraft in an operational situation wear protective suits to reflect the microwaves emitted by the high-powered radar in the aircraft.

Infrared radiation is produced by hot objects and transfers thermal energy to cooler objects. Hot objects below about 500°C produce infrared radiation only; above this temperature, visible light is also radiated. Used for thermal imaging (e.g. night-vision goggles which detect the infrared radiation given off by warm objects), remote controllers, intruder alarms, optical fibres as short wavelength infrared can carry high rates of data.

Excess infrared radiation can cause skin burns.

Visible light is a very narrow range of wavelengths that can be seen by the human eye as the colours of the visible spectrum from violet to red.

> It is used in optical fibres as light can also carry high rates of data.

Ultraviolet radiation is produced by the Sun, special ultraviolet tubes and welding arcs. The radiation can cause sunburn and skin cancer and damage eyes. It also produces vitamins in the skin and causes certain substances to fluoresce. This fluorescence can reveal markings that are invisible in visible light so is useful for security marking, detecting fake bank notes. It is used for sterilising water.

Note: the commonly used expression ultraviolet light is incorrect. Ultraviolet radiation is not part of the visible spectrum, so must *not* be called light. This misconception might occur because often ultraviolet lamps give off blue and violet light as well as ultraviolet radiation.

X-rays are produced in high-voltage X-ray tubes. They are absorbed differently by different types of matter. They can produce shadow pictures of inside the human body, which are invaluable for medical diagnosis. X-rays can penetrate inaccessible solid structures. They are used by security machines at airports and other travel hubs to scan luggage for dangerous hidden objects. X-rays are dangerous to living matter as they can kill cells and cause cell mutations which lead to cancers. Lead shielding must be used to protect people from exposure, especially those who work regularly with X-rays.

Gamma rays are produced by radioactive substances. They are very dangerous to living matter. They are used to kill cancer cells and dangerous bacteria. They are used to sterilise food and medical equipment.

Communication with electromagnetic radiation

> You should know the difference between an **analogue** and **digital signal**.
>
> You should be able to explain the benefits of digital signalling, including the increased rate of data transmission and increased range due to accurate signal regeneration.
>
> Sound can be transmitted as a digital or analogue signal.

Communication with artificial satellite is mainly by microwaves.

Some satellite phones use low-orbit artificial satellites.

Some satellite phones and television use geostationary satellites (see Topic 6 Space Physics).

Sample questions

14 State the type of orbit of satellites used for satellite television. [1]

15 State two advantages of digital signalling. [2]

Student's answers

14 Circular [0]

15 Faster and further [1]

Teacher's comments

14 Incorrect answer.

15 Partially correct answer. The signal does not travel faster. Further is a vague description but would just be enough to score a mark.

Correct answers

14 Geostationary orbit.   [1]

15 Increased rate of data transmission and increased range due to accurate signal regeneration.   [2]

Exam-style questions

Answers available at: www.hoddereducation.co.uk/cambridgeextras

13 An observatory receives X-rays and gamma rays from a star.
  a Which type of radiation has the higher wavelength?   [1]
  b Which type of radiation has the higher frequency?   [1]
  c Light from the star takes four years to reach the observatory. Do the X-rays take less, more or the same time to reach the observatory?   [1]

  d Work out the distance from the observatory to the star in metres.   [3]

14 The solid line in Figure 3.55 shows the path of a ray of blue light through a prism. A different type of radiation is directed towards the prism on the same path as the ray of blue light. The dashed line shows the path of this radiation through the prism.
  a Which type is this radiation?
    A microwaves
    B orange light
    C ultrasound
    D violet light   [1]
  b Explain your answer   [2]

▲ **Figure 3.55**

15 Which are the correct statements in the table about ultraviolet radiation?

	Its wavelength is shorter than...	A harmful effect on people of excessive exposure is...
A	visible light	skin cancer
B	visible light	burning of internal organs
C	X-rays	skin cancer
D	X-rays	burning of internal organs

                                                                        [1]

16 a Which type of electromagnetic radiation is Bluetooth?   [1]
  b Write down how there could be a problem in the transmission of Bluetooth signals within a building.   [2]

17 a i What is the nature of infrared radiation?   [1]
     ii State two uses of infrared radiation.   [2]
  b Microwaves are similar to infrared waves.
     i State one similarity and one difference between them.   [2]
     ii State a danger when people are close to a strong source of microwaves.   [1]

3.4 Sound

Key objectives

By the end of this section, you should be able to:
- know that sound waves are longitudinal waves with compressions and rarefactions produced by vibrating sources which require a medium for transmission
- be able to describe compressions and rarefactions (see 3.1 General properties of waves)
- know that the approximate range of audible frequencies to humans is 20 Hz to 20 000 Hz and that the speed of sound in air is approximately 330–350 m/s

- know that in general sound travels faster in solids than in liquids and faster in liquids than gases
- describe a method to determine the speed of sound in air involving the measurement of distance and time
- describe how changes of amplitude affect the loudness of sound waves and changes of frequency the pitch
- describe an echo as the reflection of sound waves

- describe typical uses of ultrasound

Sound waves are **longitudinal** waves that are produced by a vibrating source, which causes a material to vibrate. A material or medium is required to transmit sound waves.

The speed of sound in air at normal temperatures is 330–350 m/s.

> Sound travels faster in solids than in liquids and faster in liquids than in gases.
>
> Sound can be transmitted as a digital or analogue signal.

Although normally observed in air, sound waves can travel through liquids and solids, e.g. sea creatures communicate by sound waves travelling through water.

The healthy human ear can hear sound in air in the frequency range of 20 Hz to 20 000 Hz (20 kHz). This is called the **audible range**. In practice, only people with very good hearing can hear throughout this range. With ageing, this range is reduced and hearing tests usually only check frequencies between 250 Hz and 8 kHz.

Sound of a higher frequency than 20 kHz (the audible range) is called **ultrasound**.

> Typical uses of ultrasound:
> - medical scanning of soft tissue
> - non-destructive testing of materials
> - using sonar to determine the depth of underwater objects

The greater the amplitude of sound waves, the louder the sound.

The greater the frequency of sound waves, the higher the pitch.

Sound waves can be reflected, especially from large, hard, flat surfaces. The reflected sound is called an **echo**.

As sound travels through a material, compressions and rarefactions occur (see Figure 3.2 on p. 58). Compressions are regions where particles of material are closer together. Regions of material are rarefied where the particles move further apart (at rarefactions).

Sonar measures the time taken for ultrasound waves to be reflected from underwater objects in order to determine the depth of the object.

Skills

Determining depth from time and wave speed

An ultrasound signal transmitted to the seabed is measured to take 0.25 s to return to the ship. The speed of sound in water is 1400 m/s. Calculate the depth of the seabed.

Distance travelled by wave = speed × time = 1400 × 0.25 = 350 m

$$\text{depth of seabed} = \frac{350}{2} = 175\,\text{m}$$

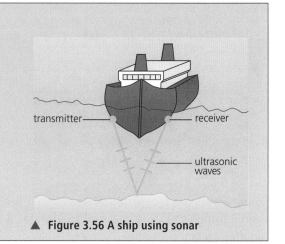

▲ **Figure 3.56 A ship using sonar**

Sample questions

16 A student stands across a field from a large building and claps their hands regularly. They hear each clap coinciding exactly with the echo from the clap before. They measure their distance from the building as 100 m and the time taken for 16 claps as 10 s. Work out the speed of sound in air. [4]

Student's answer

$\text{time between claps} = \dfrac{16}{10} = 1.6\,\text{s}$	[0]
$\text{distance travelled} = 2 \times 100 = 200\,\text{m}$	[1]
$\text{speed} = \dfrac{200}{1.6} = 125\,\text{m/s}$	[1]

Teacher's comments

The student calculated the number of claps per second instead of the time from one clap to the next. No further error was made in calculating the speed, so the student gained the last two marks.

As the source and observer are at the same place, the student correctly realised that sound travels twice the distance between the observer and the reflecting surface. Be careful in problems involving echoes not to take the distance from the observer to the reflecting surface as the distance travelled by the sound.

Unfortunately, in problems when no echo or reflection is involved, students often wrongly double the distance travelled by the sound.

Correct answer

time between claps $= \dfrac{10}{16} = 0.625\,\text{s}$ [2]

distance travelled $= 2 \times 100 = 200\,\text{m}$ [1]

speed $= \dfrac{200}{0.625} = 320\,\text{m/s}$ [1]

You must be able to describe an experiment involving the measurement of distance and time to find the speed of sound in air. (Such as this sample question.)

17 A railway worker gives a length of rail a test blow with a hammer, striking the end of the rail in the direction of its length. A sound of frequency 10 kHz travels along the rail. The speed of sound in the rail is 5000 m/s.
Calculate the wavelength of the wave. [2]

Student's answer

wavelength, $\lambda = \dfrac{v}{f} = \dfrac{3000}{10000} = 0.3\,\text{m}$ [1]

Teacher's comments

The student has used the correct equation but has substituted the wrong speed of sound.

Correct answer

wavelength, $\lambda = \dfrac{v}{f} = \dfrac{5000}{10000} = 0.5\,\text{m}$ [2]

18 A machine uses sound waves of frequency 3 MHz to form images within the human body.
Choose one of the following as the best description of these waves:
long wavelength
hypersound
polarised sound
supersonic
ultrasound [1]

Student's answer

polarised sound [0]

Teacher's comments

Incorrect response

Correct answer

ultrasound [1]

19 Dolphins emit sound waves of 95 kHz. State and explain if these waves are ultrasound. [2]

Student's answer

Sound waves of 95 kHz are ultrasound because they travel faster than normal sound waves. [1]

Teacher's comments

Ultrasound is correct but the reason is incorrect.

Correct answer

Ultrasound, because the frequency is higher than the audible range. [2]

20 At a concert attended by people of all ages, including children and old people, a sound of frequency 19 kHz is produced. Comment on how this would be heard by the audience. [1]

Student's answer

Everyone would hear it because it is in the audible range. [1]

Correct answer

People with healthy hearing would hear the sound of 19 kHz. It is right at the top of the audible range, so people with any loss of high-frequency hearing would not hear it. [1]

Teacher's comments

The answer is on the right lines but is incomplete because in such an audience it is unlikely that everyone would have completely healthy ears.

21 Is sound always transmitted as an analogue signal? [1]

Student's answer

Yes [0]

Correct answer

Sound can be transmitted as an analogue signal or a digital signal. [1]

Teacher's comments

Incorrect answer; sound is sometimes transmitted as an analogue signal but not always.

Revision activity

Sketch a sound wave by drawing vertical lines of length 1 cm to show wavefronts. Show and label compressions and rarefactions and label one wavelength.

Revision activity

Make flash cards to revise some values about sound waves. Include the speed of sound waves in air, the range of frequencies audible to humans and the frequency of ultrasound.

Exam-style questions

Answers available at: www.hoddereducation.co.uk/cambridgeextras

18 A research ship uses sonar to locate a shoal of fish. The speed of sound in water is 1500 m/s.
  a The frequency of this sound is 750 Hz.
    Calculate the wavelength of the sound waves. [2]
  b The ship receives back an echo 37 ms after a sound is transmitted.
    Work out the depth of the shoal below the ship. [3]
  c Some sound of this frequency is emitted into the air.
    i Is it audible to members of the crew of the ship?
      Explain your answer. [2]
    ii What is the wavelength of this sound in air? [2]

Key terms

Term	Definition
Alternating current (a.c.)	The direction of current flow reverses repeatedly
Direct current (d.c.)	Electrons flow in one direction only
Electric current	The charge passing a point per unit time (current $I = Q/t$ where Q is the charge flowing past a particular point in time t)
Electromagnet	Temporary magnet produced by passing an electric current through a coil of wire wound on a soft iron core
Electromagnetic induction	The production of a p.d. across a conductor when it moves through a magnetic field or is at rest in a changing magnetic field
Electromotive force (e.m.f.)	The electrical work done by a source in moving unit charge round a complete circuit
Kilowatt-hour (kWh)	The electrical energy transferred by a 1 kW appliance in 1 hour
Light-dependent resistor (LDR)	Semiconductor device in which the electrical resistance decreases when the intensity of light falling on it increases
Magnetic field	A region of space where a magnet experiences a force due to other magnets or an electric current
Magnetic materials	Materials that can be magnetised by a magnet; in their non-magnetised state, they are attracted by a magnet
Non-magnetic materials	Materials that cannot be magnetised and are not attracted to a magnet
Parallel circuit	Components connected side by side and the current splits into alternative paths and then recombines; current from the source is larger than the current in each branch
Permanent magnet	Made of steel and retains its magnetism
Potential difference (p.d.)	The work done by a unit charge passing through a component
Relay	Electromagnetic switch
Resistance	Opposition of a conductor to the flow of electric current; symbol R measured in ohms (Ω)
Series circuit	Components connected one after the other; the current is the same in each part of a series circuit
Solenoid	Long cylindrical coil of wire
Temporary magnet	Made of soft iron, and loses its magnetism easily
Thermistor	Semiconductor device in which the electrical resistance decreases when the temperature increases
Transformer	Two coils (primary and secondary) wound on a soft iron core which allow an alternating p.d. to be changed from one value to another
Conventional current	Flows from positive to negative; the flow of free electrons is from negative to positive
Electric field	A region in space where an electric charge experiences a force due to other charges
Light-emitting diode (LED)	Semiconductor device which emits light when it is forward biased but not when it is reverse biased
Potential divider	Provides a voltage that varies with the values of two resistors in series in a circuit

4.1 Simple phenomena of magnetism

Key objectives

By the end of this section, you should be able to:

- describe the forces between magnets and magnetic materials
- describe induced magnetism

  - understand that it is the interaction of magnetic fields which leads to magnetic forces

- state the differences between the properties of permanent and temporary magnets and between magnetic and non-magnetic materials
- describe a magnetic field and know how to draw the magnetic field around a bar magnet

showing its direction which is the direction experienced by a north pole at that point

- know that the spacing of magnetic field lines represents the strength of the magnetic field

- describe how to plot magnetic field lines using either iron filings or a compass and how the compass is used to determine the direction of the magnetic field.
- describe the uses of permanent magnets and electromagnets

Properties of magnets

Materials can be divided into two types:

- **magnetic materials** – materials which are attracted to magnets. Mainly the ferrous metals iron and steel and their alloys. Cobalt, nickel and certain alloys are also magnetic materials. These materials can all be magnetised to form a magnet.

- **non-magnetic materials** – materials which are not attracted to magnets and cannot be magnetised to form a magnet.

Every magnet has two poles: a north pole (N pole) and a south pole (S pole). If a magnet is supported so that it can swing freely, the N pole will always point towards the Earth's magnetic north pole. The other end is the S pole. When you bring two magnets near each other they both experience a magnetic force. If two magnets are close together, poles N and N will repel, poles S and S will repel, but poles N and S will attract.

Remember *like poles repel* and *opposite poles attract*.

Induced magnetism

A unmagnetised magnetic material can be magnetised by bringing it close to or by touching a magnet. This is called induced magnetism.
Figure 4.1 shows iron nails and steel paper clips becoming magnetised.

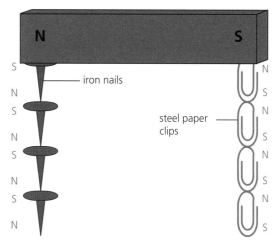

▲ **Figure 4.1 Induced magnetism**

As you can see, each nail or paperclip has their own N and S pole as they have each become magnets. If you remove the iron nails from the magnet, the chain collapses. If you take the steel paperclips away, they are still attracted to each other. This is because:

● iron is an example of a soft magnetic material (one that loses its magnetism easily and is unmagnetised easily). The induced magnetism in the iron is **temporary**. Soft magnetic materials are used to make **temporary magnets**.

● steel is an example of a hard magnetic material (one that is harder to magnetise but also harder to unmagnetise). The induced magnetism is **permanent**. Hard magnetic materials are used to make **permanent magnets**.

Magnetic fields

A **magnetic field** is a region in space where a magnet experiences a force. Figure 4.2 shows the magnetic field around a bar magnet.

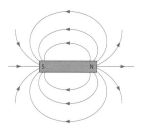

▲ **Figure 4.2 Magnetic field lines around a bar magnet**

The arrows show the direction of force. It has been agreed that the arrows show the direction of the force on the N pole of a magnet at that point.

> In Figure 4.2 you can see the magnetic field lines are closer together at the poles. This shows that the strength of the magnetic field is greater here. You can use the spacing of the magnetic field lines to work out where the magnetic field is stronger and where it is weaker.
>
> As you know, two magnets feel a force without touching. This is because their magnetic fields are interacting. It is the interaction of the magnetic fields which causes the force.

Skills

Plotting magnetic field lines

To plot magnetic field lines, you can use a compass or iron filings. To show the direction of the magnetic field you must use a compass.

Plotting compass method: The magnet is placed on a sheet of paper and a small plotting compass is placed near one pole. Mark dots on the paper at the positions of the ends of the compass needle. The compass is moved along so that the end that was over the first dot is now over the second dot. The other end is marked on the paper as the third dot. Continue this process until the other pole of the magnet is reached.

Joining the dots with a smooth line shows the field, with the direction being given by the compass arrow. Repeat for further lines starting at different points.

Iron filings method: Iron filings are sprinkled on a piece of paper placed over a magnet. When the paper is tapped gently, the filings will be seen to line up with the field lines. The field pattern can then be drawn along these lines on the paper.

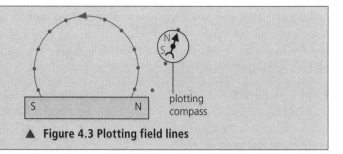

▲ **Figure 4.3 Plotting field lines**

Electromagnets

An **electromagnet** is a temporary magnet produced by passing an electric current through a coil of wire wound on a soft iron core. The soft iron core is magnetised only when there is a current in the wire.

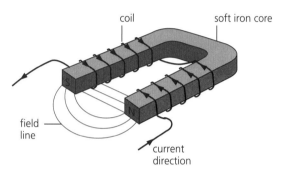

▲ **Figure 4.4 An electromagnet**

The strength of an electromagnet can be increased by:

- increasing the current
- increasing the number of coils
- moving the poles closer together

The different features of permanent magnets and electromagnets mean they have different uses. Table 4.1 summarises these.

▼ **Table 4.1 Uses of permanent magnets and electromagnets**

Type of magnet	Description	Examples of uses
Permanent	Magnetic field strength is constant and permanent	Compass, electric motor (topic 4.5.5), electric generator (topic 4.5.2), microphone, loudspeaker
Electromagnet	Temporary magnet so can be switched on and off and the magnetic field strength can be varied.	In cranes to lift scrap metal, in electric bells, magnetic locks, relays, in motors and generators.

Sample question

1 Describe how to plot the magnetic field, including its direction, around a bar magnet using iron filings. [5]

Student's answer

The iron filings should be spread around the magnet and the pattern drawn.

Use a plotting compass to find the direction. [1]

Teacher's comments

The student basically knows the right experiment, but the description is very vague and lacking in essential detail.

Correct answer

To gain full marks, the student should have given the following answer:

Place a piece of paper on top of the bar magnet. [1] Sprinkle iron filings thinly and evenly over the paper. [1] Give the paper a gentle tap. [1] Draw the field pattern on the paper along the lines of the filings. [1] Place a plotting compass on top of a field line and draw a direction arrow in the direction of the N end of the needle. Repeat to determine the directions of the whole field pattern. [1]

Exam-style questions

Answers available at: www.hoddereducation.co.uk/cambridgeextras
1 A bar magnet is suspended from its mid-point by a thread.
   a State which pole will swing towards the Earth's magnetic north pole.   [1]
   b A second bar magnet is brought close to the pole of the suspended magnet that is furthest away from the Earth's magnetic north pole. The suspended magnet does not swing away. State which pole of this magnet is closest to the suspended magnet.   [1]
2 A student is given three identical looking blocks. They are told that one is a magnet, one is made from iron and one is made from a non-magnetic material. Describe how they could use a magnet to determine which is which.   [3]
3 Cranes in scrap yards use magnets to pick up cars. Explain why an electromagnet rather than a permanent magnet is used.   [2]

Revision activity

Write all the key words from this topic on a blank piece of paper. Link the key words with lines adding explanations, e.g. you could link soft magnetic materials with electromagnet and write 'used here so that magnetism is lost when the current is switched off'. Compare your ideas with a partner to see if they have thought of any links you missed.

4.2 Electrical quantities

REVISED

4.2.1 Electric charge

Key objectives

By the end of this section, you should be able to:
● understand there are both positive and negative charges and the forces between them and explain how objects become charged by friction
● describe simple experiments to show charging by friction and how to detect charge
● describe an experiment to sort materials into electrical conductors and insulators

● recall and use a simple electron model to explain the difference between conductors and insulators

● state the unit of charge
● describe an electric field and its direction
● describe simple electric field patterns around charged objects

Positive and negative charges

When certain materials (e.g. polythene) are rubbed with a cloth, they become charged. The fact that an object is electrically charged can be detected as shown in Figure 4.5. In Figure 4.5, two polythene rods are both rubbed with a cloth. One rod is suspended freely on a thread. When the second rod is brought near, the suspended rod moves away. There is repulsion between the two charged objects. If a charged cellulose acetate rod is brought close, the suspended rod is attracted. This shows there are two types of electric charge: positive and negative. If two similarly charged objects are close together (+ and +, or − and −) they will repel, but unlike charges (+ and −) will attract.

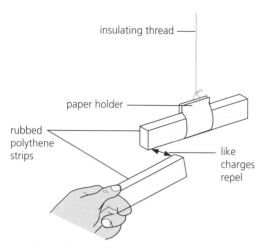

insulating thread

paper holder

rubbed polythene strips

like charges repel

▲ Figure 4.5 Investigating charges

Charges, atoms electrons

Atoms are made of small positively charged nucleus containing positively charged protons surrounded by an equal number of negatively charged electrons (Topic 5.1.1). The charge on an electron and proton is equal and opposite, and as there is the same number of each in an atom, the atom is neutral.

Rubbing an object makes it charged because the friction causes electrons to be transferred from one material to the other.

● The material gaining electrons becomes negatively charged (it now has more electrons than protons).
● The material losing the electrons becomes positively charged (it now has fewer electrons than protons).

Remember the protons are in the nucleus; they cannot move. It is only the electrons that move.

> The charge on an electron is the smallest possible quantity of charge. Charge is measured in coulombs (C). The charge on one electron is 1.6×10^{-19} coulombs, but you do not need to remember that number.

Electrons, insulators and conductors

Electrical insulators are materials in which electrons are firmly held in their atoms and cannot move, so electric charge cannot flow. Most plastics are good insulators.

Electrical conductors are materials in which electrons can move freely from atom to atom, so electric charge can flow easily. All metals and some forms of carbon are conductors.

Insulators can become charged because the charge cannot move from where the transfer happened. To charge the polythene rod (insulator) you use a cloth (another insulator). A conductor will only become charged if it is held by an insulating handle. This is because electrons are transferred between the conductor and the ground through the body of the person holding the conductor.

Skills

Detecting charges and testing materials

You can detect a charge using a gold-leaf electroscope (Figure 4.6). The gold leaf is only attached to the metal rod by the top edge and is free to move. The rod is held in place by a plastic support and in a wooden box with glass sides to protect from any movement of air.

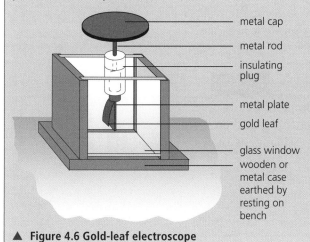

To detect a charge: Bring the charged rod near to the metal cap of the electroscope. The gold leaf rises away from the plate. When the charged rod is removed, the leaf falls. The gold leaf will behave the same whether the object is negatively or positively charged.

Charging the electroscope: Drag a charged polythene rod (negatively charged) across the surface of the metal cap. The leaf will rise and remain up even when the rod is removed. Electrons have been transferred from the polythene rod to the electroscope. The electroscope is now negatively charged.

Testing materials to see if they are insulators or conductors: Touch the charged electroscope with different objects made of different materials, e.g. a wooden or glass rod, or a piece of metal or plastic. If the leaf falls, the object is a conductor. If the leaf remains in place, the object is an insulator.

▲ **Figure 4.6 Gold-leaf electroscope**

Electric field

An **electric field** is a region in space where an electric charge experiences a force. The field can be represented by lines of force in the direction that a positive charge would move if placed at a point in the field. You should be able to describe the pattern and direction of the electric field around a point charge (Figure 4.7a), a charged conducting sphere (Figure 4.7b) and between two oppositely charged parallel plates (Figure 4.7c).

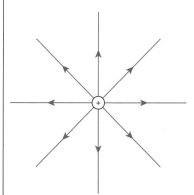

▲ **Figure 4.7a Electric field around a positive point charge**

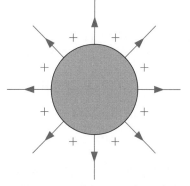

▲ **Figure 4.7b Electric field around a positively charged sphere. Notice the field lines appear to start in the centre of the sphere. Do not draw them inside though.**

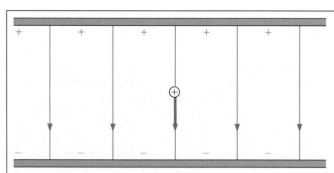

▲ **Figure 4.7c Electric field between two parallel plates. This is a uniform field. You can tell because the field lines are parallel and evenly spaced.**

Sample question

2 An inkjet printer produces a stream of very small droplets from a nozzle. The droplets are given a negative electric charge and then pass between two plates with positive and negative charge, as shown in Figure 4.8.

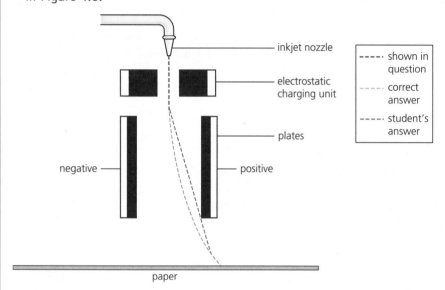

▲ **Figure 4.8**

    a State the type of field in the region between the charged plates. [1]
    b State and explain the force acting on the droplets in this region. [2]
    c Extend the dashed line to complete the path of the droplets. [3]

Student's answers

    a *There is an electric field in this region.* [1]
    b *There is a force to the right on the ink droplets.* [1]
    c *The student's answer is shown by the red dashed line in Figure 4.8.* [2]

Correct answers

    a There is an electric field in this region. [1]
    b There is a force to the right on the ink droplets, which are attracted to the positive plate. [2]
    c The correct answer is shown by the blue dashed line in Figure 4.8. [3]

Teacher's comments

    a Correct answer.
    b The student correctly stated that the force is to the right but failed to explain it.
    c The path between the plates must be curved, as the droplets are accelerated to the right by the electric force.

4.2.2 Electric current

Electric current is a flow of charge. In metals there are free electrons. These are electrons that are only loosely attached to a particular atom. When you connect a battery across the ends of a metal wire, these free electrons start to move. They move slowly in the direction of the positive terminal of the battery (remember positive and negative charges attract).

Calculating current

An electric current is defined as the charge passing a point per unit time. This can be written as:

$$I = \frac{Q}{t}$$

where I is the current and Q is the charge flowing past a particular point in time t.

Conventional current

The agreed direction of **conventional current** in a circuit is from the positive terminal to the negative terminal of a battery. The electrons flow in the opposite direction to conventional current.

Ammeters

Electric current is measured in amperes, usually abbreviated to amps (symbol A), by an ammeter, which must be connected in **series**. The positive terminal of the ammeter is connected to the positive terminal of the supply. Figure 4.9 shows an ammeter in series with component X.

Ammeters can be analogue (have a scale and pointer like a moving-coil ammeter) or have a digital display. Whichever type you use you should use the correct range. If the current you are measuring is unknown, it is good practice to start on the largest scale, for example, 0 to 10 A. If the reading is very small or not detected, you can reduce the range perhaps to 0 to 1 A. Choose the smallest possible range to get the most accurate result, e.g if the current is 0.02 A, you would choose the 100 mA range.

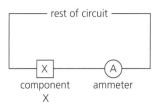

▲ **Figure 4.9 Measuring the current through component X**

Skills

Expressing quantities using multipliers

You should be able to use the common multipliers. These are used to express very large or very small numbers.

▼ **Table 4.2 Common prefixes**

Prefix	Symbol	Multiply by	Prefix	Symbol	Multiply by
centi	c	$\times 10^2$			
kilo	k	$\times 10^3$	milli	m	$\times 10^{-3}$
mega	M	$\times 10^6$	micro	μ	$\times 10^{-6}$
giga	G	$\times 10^6$	nano	n	$\times 10^{-9}$

For example,

$1200\,000\,N = 1.2\,MN$ $0.03\,A = 30\,mA$ $5\,km = 5000\,m$

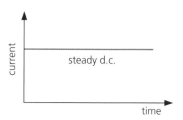

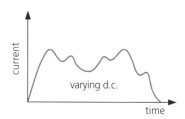

▲ **Figure 4.10a Direct current**

Direct and alternating current

In a **direct current (d.c.)**, the charge flows in one direction only (Figure 4.10a).

In an **alternating current (a.c.)**, the direction of the current changes repeatedly (Figure 4.10b).

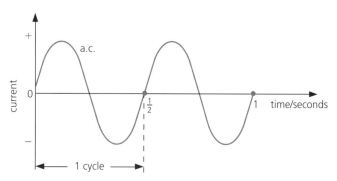

▲ **Figure 4.10b Alternating current**

The number of complete cycles per second of a.c. is the frequency of the alternating current.

Sample question

REVISED

3 A charge of 35 C flows around a circuit in 14 s.
   a Calculate the current flowing. [2]
   b The charge on each electron is 1.6×10^{-19} C. Calculate the number of electrons flowing around the circuit in this time. [2]

Student's answers

a $I = \dfrac{Q}{t} = \dfrac{35}{14} = 2.5\,A$ [2]

b $number\ of\ electrons = \dfrac{35}{1.6 \times 10^{-19}} = 2.19 \times 10^{19}$ [1]

Teacher's comments

a Correct answer.
b The student started the calculation correctly, but made a mistake in the calculation of the powers of 10.

Correct answers

a $I = \dfrac{Q}{t} = \dfrac{35}{14} = 2.5\,\text{A}$ [2]

b number of electrons $= \dfrac{35}{1.6 \times 10^{-19}} = 2.2 \times 10^{20}$ to 2 s.f. [2]

Revision activity

There are three important electrical quantities used to describe circuits that are sometimes confused: current (Topic 4.2.2), electromotive force (Topic 4.2.3) and resistance (Topic 4.2.4). To help you remember and understand them, take a piece of plain paper and divide it into three equal sections. In the first section, summarise the information about electric current.

4.2.3 Electromotive force and potential difference

Key objectives

By the end of this section, you should be able to:
- define electromotive force (e.m.f.) and potential difference (p.d.) and know that they are both measured in volts
- describe the use of voltmeters
- recall and use the correct equations for e.m.f. and p.d.

Electromotive force (e.m.f.) is the electrical *work done by a source* in moving unit charge around a circuit. It is measured in volts.

Potential difference (p.d.) across a component is the *work done by unit charge* passing through a component. It is also measured in volts.

Calculating e.m.f. and p.d.
The equations for e.m.f (E) and p.d. (V) appear the same:

$$E = \frac{W}{Q} \text{ and } V = \frac{W}{Q}$$

When calculating e.m.f. (E), W is the energy transferred to an amount of charge Q by the battery or power supply.

When calculating p.d. (V), W is the work done when the amount of charge Q passes between two points.

Voltmeters

The p.d. across a component is measured by a voltmeter, which must be connected in **parallel** with the component. It can be helpful to build the circuit and then place the voltmeter across the component. The positive terminal of the voltmeter should be connected to the side of the component nearest to the positive side of the battery. Figure 4.11 shows a voltmeter in parallel with component X.

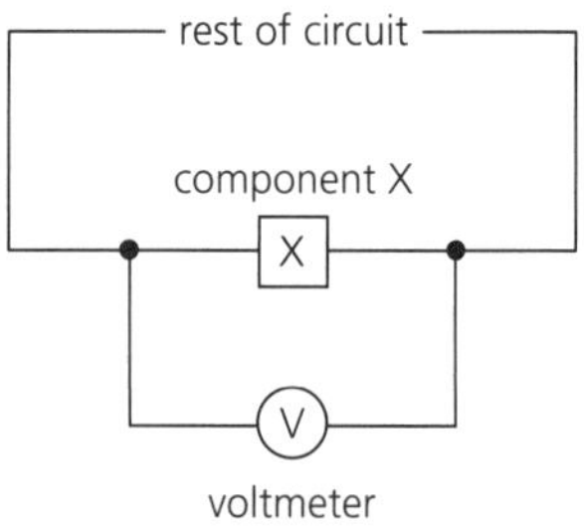

Voltmeters can be analogue (have a scale and pointer as in Figure 4.12) or have a digital display.

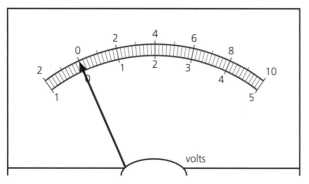

▲ **Figure 4.12 An analogue voltmeter scale**

Whichever type you use, you should use the correct range. Choose the smallest possible range to get the most accurate result. For example, if you expect the measurement to be 12 V, you must choose a suitable scale such as 0 to 20 V. If the reading is 0.006 V, you should use 0 to 10 mV.

▲ **Figure 4.11 Measuring the p.d. across component X**

Revision activity

In the second section of your electrical quantities sheet, summarise the information about e.m.f and p.d.

4.2.4 Resistance

Key objectives

By the end of this section, you should be able to:
- recall and use the correct equation to calculate resistance
- describe an experiment using a voltmeter and ammeter to determine the resistance of a component using the correct equation
- describe how resistance of a wire is affected by its length and its cross-sectional area
- sketch and explain the voltage–current graphs for a resistor, a filament lamp and a diode
- recall and use the relationship between the resistance of a wire and its length and cross-sectional area

Resistance is the opposition of a conductor to the flow of electric current.

Resistance is calculated using the equation:

$$R = \frac{V}{I}$$

where R = resistance in ohms (Ω), V = p.d. and I = current.

Skills

Determining resistance

To determine the resistance of an unknown component, you must connect the circuit as shown in Figure 4.13.

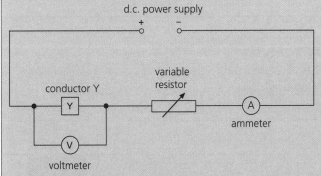

d.c. power supply

conductor Y

variable resistor

ammeter

voltmeter

▲ **Figure 4.13 Circuit to determine the resistance of unknown conductor Y**

1 The ammeter is connected in **series** with the power supply and conductor Y (so that the current flows *through* the ammeter).
2 The voltmeter is in **parallel** with conductor Y (so that the voltmeter measures the p.d. *between* the ends of the conductor).
3 Record the values of the p.d. in volts and the current in amps.
4 Change the setting of the variable resistor and record at least five more pairs of values.
5 Work out the value of the resistance using the equation $R = \dfrac{V}{I}$ for each pair of readings.

Current–voltage graphs

If you plot the values of current and voltage from your experiment on a graph, you get distinctive shape graphs for different components (Figure 4.14).

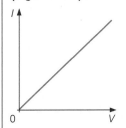

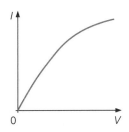

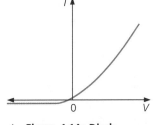

▲ **Figure 4.14a Resistor** ▲ **Figure 4.14b Filament lamp** ▲ **Figure 4.14c Diode**

The resistance is equal to the inverse of the gradient of the current–voltage graph of a component.

A resistor or metal wire at constant temperature shows a directly proportional relationship (a straight line through the origin). The gradient is constant because the resistance is constant (Figure 4.14a).

For a filament lamp, the gradient decreases. This shows the resistance is increasing. As the voltage and current increase, the temperature of the filament increases. Resistance increases with a large increase in temperature (Figure 4.14b).

Diodes (including LEDs) have an extremely high resistance in the reverse direction, so no reverse current can flow. Resistance is very low in the forward direction, so current can flow easily (Figure 4.14c). Diodes are used in circuits to make sure the current only flows in one direction. They can be used to change alternating current to direct current.

Resistance of a metal wire

The resistance of a metal wire depends on:

● length – the longer the wire, the greater the resistance
● cross-sectional area – the thicker the wire, the smaller the resistance
● the material the wire is made from

Resistance is directly proportional to the length. This means if you double the length, the resistance will double. Resistance is inversely proportional the cross-sectional area. This means if you double the cross-sectional area, the resistance halves.

Sample questions

REVISED

4 A student carries out an experiment to find the resistance of a wire. They vary the supply voltage and take measurements using an ammeter and a voltmeter. The table shows their first three readings.

Reading	1	2	3	4
voltage/V	1.10	2.10	2.95	1.75
current/A	0.20	0.35	0.50	?
resistance/Ω	5.50	6.00	?	?

Make sure you show your working for all parts of this question.

a Calculate the resistance for the third pair of readings. [1]
b Calculate the average value of resistance for the first three readings. [1]
c Using this average value of resistance, calculate the current the student can expect when they take the fourth reading. [2]

Student's answers

a Reading 3: $R = \dfrac{V}{I} = \dfrac{2.95}{0.50} = 5.9\,\Omega$ [1]

b Average $V = \dfrac{1.10 + 2.10 + 2.95}{3} = \dfrac{6.15}{3} = 2.05\,V$

Average $I = \dfrac{(0.20 + 0.35 + 0.50)}{3} = \dfrac{1.05}{3} = 0.35\,A$

Average $R = \dfrac{2.05}{0.35} = 5.86\,\Omega$ [0]

c Reading 4: $I = VR = 1.75 \times 5.86 = 10.26\,A$ [0]

Teacher's comments

a Correct answer.
b The student should have taken the average of the three values of resistance, using the answer for part a. It is simply good fortune that their answer is close to the correct answer.
c The student has rearranged the equation $V = IR$ incorrectly. Observation and comparison with the first three readings show that the calculated value is unlikely.

Correct answers

a Reading 3: $R = \dfrac{V}{I} = \dfrac{2.95}{0.50} = 5.9\ \Omega$ [1]

b Average $R = \dfrac{5.50 + 6.00 + 5.90}{3} = \dfrac{17.40}{3} = 5.8\ \Omega$ [1]

c Reading 4: $I = \dfrac{V}{R} = \dfrac{1.75}{5.80} = 0.30$ A to 2 s.f. $\left(0.3017\ \text{A}\right)$ [2]

5 Sample A is a length of wire of given material.
   a Copy and complete the table for the resistance of three more samples of wire of the same material. Choose from the following words: greater, less, same.

Sample	B	C	D
length compared with A	×2	same	×2
diameter compared with A	same	$\dfrac{1}{2}$	$\dfrac{1}{2}$
resistance compared with A			

[3]

   b Add numerical values to your entries in the table to show the magnitude of resistance compared with sample A. [3]

Student's answers

Sample	B		C		D	
length compared with A	×2		same		×2	
diameter compared with A	same		$\dfrac{1}{2}$		$\dfrac{1}{2}$	
resistance compared with A	*greater ×2*	[2]	*greater ×2*	[2]	*greater ×4*	[1]

Teacher's comments

a Correct answers – all three samples have greater resistance than sample A.

b The resistance of sample B will be ×2 greater – the student's answer to B is correct. Resistance varies with the inverse of area, not diameter, so the answer to C is incorrect. Although the answer to D is incorrect, the student has correctly carried over from the answer to C, so no further marks are lost. The answer for C should be ×4 and the answer for D should be ×8.

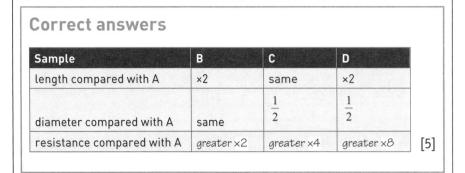

Sample	B	C	D
length compared with A	×2	same	×2
diameter compared with A	same	$\frac{1}{2}$	$\frac{1}{2}$
resistance compared with A	greater ×2	greater ×4	greater ×8

[5]

4.2.5 Electrical working

Key objectives

By the end of this section, you should be able to:
- understand that a cell or mains supply is a source of electrical energy and that circuits transfer this energy to the circuit components and then to the surroundings
- recall and use the correct equations for electrical power and electrical energy
- define the kilowatt-hour and calculate the cost of using electrical appliances

Electrical energy and electrical power

The electrical cell, battery or mains supply are a source of electrical energy. Electric circuits transfer this energy to components in the circuit and then into the surroundings. For example, a torch battery transfers energy to the lamp; this energy is then transferred by light and by heating to the surroundings.

To calculate the electrical energy transferred, use the equation:

$E = Ivt$

where E is the energy transferred (J), I is the current (A), V is the p.d. (V) and t is the time (s).

It is also useful to consider the electrical power of an appliance. Remember power is the amount of energy transferred every second ($P = W/t$); the unit of power is the watt (W).

To calculate electrical power, use the equation:

$P = IV$

Paying for electricity

When buying electricity from a supply company, a much larger unit of electricity is used: the **kilowatt-hour (kW h)**. This is the electrical energy transferred by a 1 kW appliance in 1 hour. The company then charges a price per kilowatt–hour for the energy transferred. Remember 1 kW is 1000 W.

To calculate the energy transferred in kW h, use the equation:

energy transferred (kW h) = power (kW) × time (h)

Sample question

REVISED

6 A travel kettle is designed for international use. With a 230 V supply, the power rating is 800 W.
  a Calculate the current with a 230 V supply and the resistance of the element. [2]

b Find the current and power output of the kettle when used in another country with a 110 V supply. [3]
c Comment on the use of this kettle in the country with the 110 V supply. [1]

Student's answers

a $I = \dfrac{P}{V} = \dfrac{800}{230} = 3.48\,A$ $R = \dfrac{V}{I} = \dfrac{230}{3.48} = 66.1\Omega$ [2]

b I will stay the same.
$P = 110 \times 348 = 383\,W$ [1]

c The kettle will take longer to boil water. [1]

Correct answers

a $I = \dfrac{P}{V} = \dfrac{800}{230} = 3.48\,A$

$R = \dfrac{V}{I} = \dfrac{230}{3.48} = 66.1\,\Omega$ [2]

b The element is the same so R will stay the same.

$I = \dfrac{110}{66.1} = 1.66\,A$

$P = 110 \times 1.67 = 183\,W$ [3]

c The kettle will take much longer to boil water. [1]

Teacher's comments

a Correct answers.
b The current cannot stay the same because R is the same but V is different. 1 mark was awarded as the student calculated the power from the wrong value of current without any further error.
c The student made a valid comment from the calculated value of P.

Exam-style questions

Answers available at: www.hoddereducation.co.uk/cambridgeextras
4 When rubbed with a dry cloth, Perspex becomes positively charged. Polythene becomes negatively charged when rubbed with a dry cloth
  a Describe an experiment you could do to show the two rods have opposite charges. [3]
  b Explain in terms of electron movement how a Perspex rod becomes positively charged when rubbed with a dry cloth. [3]
  c Each time the cloth is used it also becomes charged. State and explain the charge on the cloth when it is used to charge the polythene rod. [2]

5 Describe the electric field around a negative point charge. [2]

6 A student is attempting to measure the current through a component. They are told it will be approximately 0.15 A. Choose the most appropriate range for the ammeter. [1]
0 to 10 A   0 to 1 A   0 to 100 mA

7 A charge of 75 C passes a point in the circuit in 5 minutes. Calculate the current in the circuit. [2]

8 Use the analogue voltmeter in Figure 4.12 to answer this question.
  a Name the two ranges shown on the voltmeter. [2]
  b State the value of the small divisions between 4 and 6 on the voltmeter. [1]
  c A student connects the voltmeter into their circuit and the needle moves to the left. Describe what they should do to correct this. [1]

Revision activity

Create flash cards for all the equations and key words in Topic 4.2 and learn them.

9 You are asked to take measurements from the circuit shown in Figure 4.15 and are provided with an ammeter, a voltmeter and any necessary connecting wires.

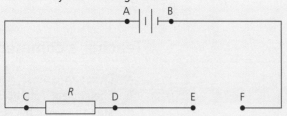

▲ **Figure 4.15**

Complete the table to indicate which component, if any, you should connect across the points AB, CD and EF to take each measurement. [2]

Measurement to be taken	AB	CD	EF
current through *R* when connected to battery			
p.d. across *r* when connected to battery			

10 a Calculate the e.m.f. of a battery if it does 45 J of work moving 15 C of charge around a complete circuit. [1]

b The p.d. across a lamp is 12 V. Calculate the energy transferred to the lamp if 180 C passes through it. [2]

11 A lamp has a potential difference of 6.0 V across it and a current of 1.5 A through it. Calculate its resistance. [1]

12 The potential difference across a resistor is 2.5 V and it has a resistance of 20 Ω. Calculate the current in the resistor. [2]

13 A cylindrical block of conducting putty has a resistance of 100 Ω. Explain a change you could make to the block to reduce the resistance to 25 Ω. [4]

14 A microwave has a power rating of 800 W. It is used for 30 minutes during the day. Calculate the energy transferred by the microwave in the day in kW h. [3]

15 The potential difference across a lamp is 12 V and the current is 4.0 A. The lamp is on for 4 minutes.
Calculate:
a the energy transferred by the lamp in joules [2]
b the power of the lamp [1]

16 A 3 kW electric heater is used to heat up 2.5 kg of water of specific heat capacity 4200 J/(kg°C). The initial water temperature is 16°C. Calculate the temperature of the water after the heater has been switched on for 2 minutes. [4]

4.3 Electric circuits

REVISED

4.3.1 Circuit diagrams and components

Key objectives

By the end of this section, you should be able to:
● draw and understand circuit diagrams containing many different circuit components including diodes and light-emitting diodes (LEDs)

You must be able to draw and interpret all the symbols shown in Figure 4.16.

> ## Revision activity
> Test yourself by creating flash cards of the circuit symbols. Make sure you recognise each of them and can draw each of them.

cell
battery of cells
battery of cells
power supply
a.c. power supply
fixed resistor
variable resistor
heater
light-dependent resistor
thermistor
ammeter
A
voltmeter
V
generator
G
lamp
switch
transformer
magnetising coil
relay switch
motor
M
fuse
potential divider
junction of conductors
earth or ground
diode
light-emitting diode

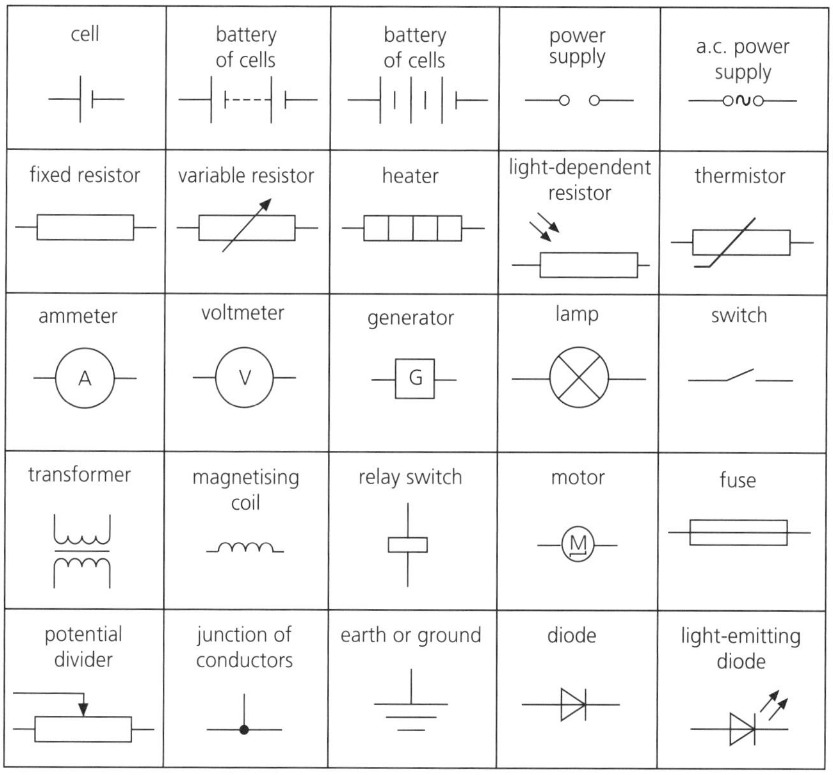

▲ **Figure 4.16 Circuit symbols**

4.3.2 Series and parallel circuits

Key objectives

By the end of this section, you should be able to:

- understand current and potential difference in a series circuit
- understand how to build and use series and parallel circuits
- calculate the combined e.m.f. and combined resistance for components in series
- understand current and resistance in a parallel circuit
- state the advantages of using parallel circuits for lighting
- calculate current, potential difference and resistance for components in parallel
- explain why the sum of the currents entering a junction equals the sum of the currents leaving the junction

Series circuits

In a **series circuit**, there is just one path for the current to follow.

V
R1
R2
R3
I
I
V1
V2
V3

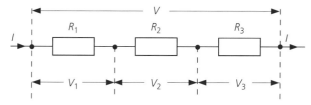

▲ **Figure 4.17 Resistors in series**

Rules for components in series:

- The current at every point in a series circuit is the same, I.

● The total resistance (R_T) in a series circuit is the sum of the individual resistances:

$$R_T = R_1 + R_2 + R_3$$

● The potential difference across components in series is equal to the sum of p.d.s across each component:

$$V = V_1 + V_2 + V_3$$

● The combined e.m.f. of different sources in series is the sum of each individual e.m.f. For example, if you connect two cells which each have an e.m.f. of 1.5 V, the total e.m.f. is 3.0 V.

Parallel circuits

In a **parallel circuit**, there are alternative paths or branches for the current.

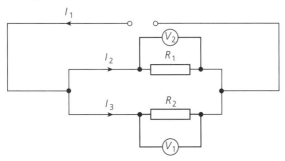

▲ **Figure 4.18 Resistors in parallel**

Rules for components in parallel:

● When components are in parallel, the current from the source is greater than the current in each branch.

● The combined resistance of components in parallel is less than the resistance of any one resistor.

● Sum of the currents entering a junction = sum of the currents leaving the junction:

$$I_1 = I_2 + I_3$$

● The p.d. across components in parallel is the same:

$$V_1 = V_2$$

● The total resistance R_T of R_1 and R_2 is given by:

$$\frac{1}{R_T} = \frac{1}{R_1} + \frac{1}{R_2}$$

Current is a flow of electrons. The electrons cannot be created or destroyed. The same number must flow per second through every point in a circuit. This is why the sum of the currents entering a junction is equal to the sum of the currents leaving the junction. It is also why the current in a series circuit is the same at every point.

In lighting circuits in homes and businesses, lamps are connected in parallel. This is because:

● Each lamp has the same p.d. across it, which is the p.d. of the supply. Each lamp is therefore the same brightness and you can have as many lamps as you want in the circuit.

● You can switch each lamp on and off individually. If one lamp should fail, the other lamps will continue to work.

Sample questions

7 In Figure 4.17, $R_1 = 4\,\Omega$ and $R_2 = 3\,\Omega$.
    a Calculate the total resistance of R_1 and R_2.   [2]
    b The current through R_1 is 1.5 A. State the current through R_2.   [2]
    c Calculate the potential differences V_1 and V_2.   [2]

> **d** The potential difference across all three resistors is 12 V.
> Calculate the value of V_3 and hence the resistance of R_3.   [4]

Student's answers

> *a*   total resistance $= 4 + 3 = 7\,\Omega$ [2]
>
> *b*   current through $R_2 = \dfrac{3}{4} \times$ current through $R_1 = 0.75 \times 1.5 = 1.125\,\text{A}$ [0]
>
> *c* $V_1 = I_1 \times R_1 = 1.5 \times 4 = 6\,\text{V}$
> $V_2 = I_2 \times R_2 = 1.125 \times 3 = 3.375\,\text{V}$ [2]
>
> > *d* $V_3 = 12 - (4 + 3) = 12 - 7 = 5\,\text{V}$
> >
> > $R_3 = \dfrac{V_3}{I_3} = \dfrac{5}{1.125} = 4\,\Omega$ [2]

Teacher's comments

> **a** Correct answer
> **b** The student did not recognise that current stays the same through components in series.
> **c** The student correctly applied their answers from part **b**.

> **d** The student might have been on the right lines and then made errors in substituting resistances instead of voltages. With little working and no explanation, it is impossible for the examiner to know. The answer for R_3 followed on reasonably from earlier working, so credit was gained for this.   [2 marks given]

Correct answers

> *a*   total resistance $= 4 + 3 = 7\,\Omega$ [2]
> *b*   current through $R_2 =$ current through $R_1 = 1.5\,\text{A}$ [2]
> *c* $V_1 = I \times R_1 = 1.5 \times 4 = 6\,\text{V}$
> $V_2 = I \times R_2 = 1.5 \times 3 = 4.5\,\text{V}$ [2]
>
> > *d*   supply potential difference = sum of p.d.s of rest of circuit = 12 V
> > $12 = V_1 + V_2 + V_3$
> > $12 = 6 + 4.5 + V_3$
> > $V_3 = 12 - (6 + 4.5) = 12 - 10.5 = 1.5\,\text{V}$
> > $R_3 = \dfrac{V_3}{I} = \dfrac{1.5}{1.5} = 1\,\Omega$ [4]

8 In Figure 4.18, $R_1 = 4\,\Omega$, $R_2 = 3\,\Omega$, $I_1 = 4.2$ A and $I_2 = 1.8$ A.
   a Calculate the current I_3. [2]
   b Calculate the total resistance of R_1 and R_2. [2]
   c Calculate the e.m.f. of the power supply. [2]

Student's answers

a current $I_3 = I_1 + I_2 = 4.2 + 1.8 = 6.0$ A [0]

b total resistance $= \dfrac{1}{R_1} + \dfrac{1}{R_2} = \dfrac{1}{4} + \dfrac{1}{3} = 0.25 + 0.333 = 0.583\ \Omega$ [0]

c e.m.f. $= 6.0 \times 0.583 = 3.50$ V [2]

Teacher's comments

a The student has wrongly thought that I_3 is the total current.
b The student has applied the wrong equation.
c Full marks are given despite the wrong answer. The student correctly followed on from parts a and b.

Correct answers

a current $I_3 = I_1 - I_2 = 4.2 - 1.8 = 2.4$ A [2]

b $\dfrac{1}{R_T} = \dfrac{1}{R_1} + \dfrac{1}{R_2} = \dfrac{1}{4} + \dfrac{1}{3} = \dfrac{3+4}{12} = \dfrac{7}{12}$, $R_T = 1.7\ \Omega$ [2]

c e.m.f. $= 2.4 \times 3 = 7.2$ V [2]

Revision activity

Create a table summarising the key features of series and parallel circuits. Include circuits diagrams.

4.3.3 Action and use of circuit components

Key objectives

By the end of this section, you should be able to:
● understand that for a constant current, the p.d. across a conductor increases as its resistance increases

● describe how a variable potential divider can change the voltage output, and recall and use the correct equation for two resistors used in a potential divider.

Increase in resistance of a conductor

Consider a conductor with a constant current through it. The product of the resistance and the current gives the potential difference ($V = IR$, remember the resistance $R = V/I$).

If the resistance of the conductor increases and the current remains constant, then the potential difference across the conductor increases.

Light-dependent resistors and thermistors

The resistance of a **light-dependent resistor (LDR)** falls with increasing light level. It can be connected in a circuit that is required to respond to changes in light level.

The resistance of a **thermistor** decreases considerably with increasing temperature.

These components can be used as an input to a switching circuit such as in a security light that only works at night or in a fire alarm that switches on when it detects heat.

Relays

Switching circuits cannot power the appliance they are switching on, for example, starting the motor of a washing machine when the water is at the correct temperature. A **relay** is a switch turned on or off by an electromagnet.

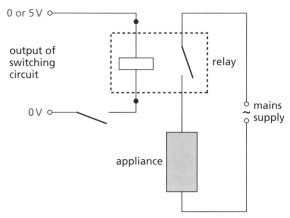

▲ **Figure 4.19 A relay is used to switch on a mains appliance**

The small current from the switching circuit, switches on the electromagnet which attracts the switch, closing it. The appliance is then switched on.

Variable potential divider

A **potential divider** provides a voltage that varies with the values of two resistors in series in a circuit. Figure 4.20 shows a potential divider with two separate resistors R_1 and R_2.

If the value of R_1 or R_2 changes, the output voltage will change.

If R_1 increases with R_2 unchanged, V_1 increases and V_2 decreases.

If R_2 increases with R_1 unchanged, V_2 increases and V_1 decreases.

Remember the sum of p.d.s for components in series is equal to the total p.d. The ratio of the voltages across the two resistors is given by:

$$\frac{R_1}{R_2} = \frac{V_1}{V_2}$$

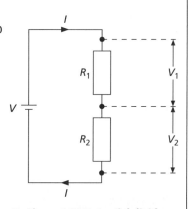

▲ **Figure 4.20 Potential divider with two separate resistors**

Skills

Using ratios to determine the p.d. across each resistor in a potential divider arrangement

You can use this ratio to determine the p.d. across each resistor.

For example, in Figure 4.20, $R_1 = 20\,\Omega$ and $R_2 = 80\,\Omega$.

First, find the ratio of the resistors:

$$\frac{20}{80} = \frac{1}{4}, \text{ so ratio is } 1{:}4$$

Next, add together the parts of the ratio to find the total number of shares:

number of shares $= 1 + 4 = 5$

Then multiply the supply voltage by the proportion of shares required:

$$V_1 = 1 \times \frac{V}{5} \text{ and } V_2 = 4 \times \frac{V}{5}$$

Looking at another example: two resistors R_1 and R_2 are in series with an e.m.f. $= 12\,V$. $R_1 = 10\,\Omega$ and $R_2 = 50\,\Omega$. Calculate the p.d. across R_2.

Ratio of resistors:

$$\frac{10}{50} = \frac{1}{5}, \text{ so ratio is } 1{:}5$$

number of shares $= 1 + 5 = 6$

$$V_2 = 5 \times \frac{12}{6} = 10\,V$$

Figure 4.21 shows a circuit that acts as a fire alarm. When the temperature of the thermistor rises, its resistance falls. The thermistor and fixed resistor R are a potential divider, so the p.d. between S and T rises and enough current flows into the relay for it to switch on the bell.

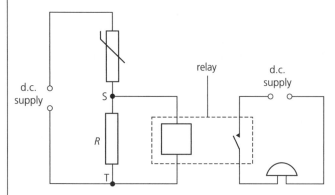

▲ **Figure 4.21 Fire alarm circuit**

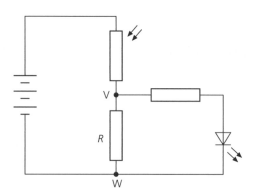

▲ **Figure 4.22 Light-sensitive circuit using a remote light-emitting diode**

Figure 4.22 shows a circuit that acts as a warning when too much light enters an automated photographic laboratory. A **light-emitting diode (LED)** on the control panel outside the laboratory can light up to show the warning.

When operating correctly in the dark, the resistance of the LDR is high. The p.d. between V and W is low, so no current flows through the LED.

If light enters the laboratory, there is an increase in light level and the resistance of the LDR falls. The LDR and fixed resistor R are a potential divider, so the p.d. between V and W rises and enough current flows through the LED for it to light up and give a warning.

Light-emitting diodes (LED)
LEDs will only light when forward biased (current in direction of the arrow). LEDs must have a resistor in series with it to limit the current as diodes have low resistance in forward bias and can easily be damaged. They are very useful as indicator lights.

Exam-style questions

Answers available at: www.hoddereducation.co.uk/cambridgeextras

17 A student has three 1.5 V cells. Describe how they could combine them to have a p.d. of 4.5 V for their circuit. [2]

18 A student has a 20 Ω resistor and a 40 Ω resistor.

　a Calculate their combined resistance if they are connected in series. [1]

　b Calculate their combined resistance if they are connected in parallel. [2]

19 In Figure 4.18, the current $I_1 = 0.5$ A, the current $I_2 = 0.3$ A and the resistance of $R_1 = 40 \, \Omega$.

　a Calculate the p.d. across R_1. [2]

　b Calculate the resistance of resistor R_2. [3]

20 The supply voltage in Figure 4.20 is 12 V and $R_1 = 25 \, \Omega$ and $R_2 = 75 \, \Omega$. Calculate the potential difference across each resistor. [3]

21 Figure 4.23 shows a circuit. Describe what happens when the temperature increases. [4]

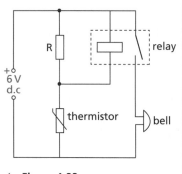

▲ **Figure 4.23**

4.4 Electrical safety

REVISED

Key objectives

By the end of this section, you should be able to:

- state possible hazards when using mains electricity
- understand that the mains circuit consists of three wires – live, neutral and earth – and explain why switches are placed in the live wire
- explain how trip switches and fuses work and choose appropriate settings and values for each
- explain why metal outer casings for electrical appliances are earthed
- state that for double-insulated appliances, the fuse protects the appliance from current surges even without the earth

Dangers of electricity

Some common hazards when using a mains supply are:

- Damaged insulation can lead to very high currents flowing in inappropriate places. This poses a danger of electric shock or fire.
- Cables that overheat owing to excessive current can lead to fire or damage in the appliance or in the cables and insulation.
- Damp conditions. Water lowers the resistance to earth, so damp conditions can lead to the current shorting and can cause shocks. Electrical devices for use in damp conditions must be designed to high standards of damp proofing, especially connectors and switches.
- Overloading plugs, extension leads or sockets. If you have too many appliances plugged into one outlet, then the current will be too great. This can cause overheating and so is a fire risk.

House circuits

Mains circuit wire consists of three wires: the live wire, the neutral wire and the earth wire. The mains supply is usually a.c. and the potential difference of the live wire with respect to earth varies depending on your country (lowest value of any country is 110 V a.c. and highest value is 240 V a.c.). The neutral and earth wires are at 0 V with respect to earth.

The circuits in a home are all connected in parallel across the live and neutral wires. This means they all receive the mains p.d. and can be switched on and off separately. The switch is placed in the live wire, so that when the appliance is switched off it is not connected to the mains supply.

Fuses and trip switches

A fuse is a piece of wire that melts and breaks when too much current flows through it. This switches off the circuit to protect against shock, fire or further damage. Fuses are placed in the live wire to safely switch off the device.

Trip switches or circuit breakers contain electromagnets which when the current is large enough will separate contacts and break the circuit. They operate quickly and can be easily reset by pressing a button.

Fuses come in different values and trip switches have to be set to the right setting. To choose the correct values, you need to know the maximum current expected. For example, if the maximum current is 9 A you might use a 13 A fuse or choose a 10 A trip switch.

Earthing

Appliances with metal cases have to be earthed using the earth wire. This is to protect against electric shock. For example, if the live wire became loose and touched the metal casing, the whole appliance would become live. To prevent this happening, the earth wire is connected to the metal appliance and connected to earth. If the case became live, a large current would pass through the live wire to earth through the earth wire. The fuse will melt and break, switching off the appliance.

Double insulation

Many electrical appliances have a plastic outer case, they are double-insulated. As plastic is an insulator, there is no risk of shock and these appliances do not need an earth connection. They will still have a fuse as it protects the appliance from current surges due to a short circuit.

Sample question

REVISED

9 People are gathered after dark on wet grass. Explain whether the following three situations are potentially dangerous:
  a A heater and several high-powered electric lamps are supplied by an old extension cable. [2]
  b There is a cut in the outer insulation of the cable. [2]
  c The devices are connected to a switch lying on the lawn. [2]

Student's answers

  a The electrical power is likely to require too much current in the cable, leading to overheating. This could cause a fire or melting of the insulation. [2]
  b There is insulation on the individual wires, so the cable is safe. [0]
  c The dew on the cable connection could cause an electric shock. [2]

Teacher's comments

  a Correct answer
  b Incorrect answer – using a cable with any sort of cut is unsafe practice.
  c Correct answer.

Correct answers

  a The electrical power is likely to require too much current in the cable, leading to overheating. This could cause a fire or melting of the insulation. [2]
  b There could also be a cut in the insulation of the individual wires, which would be difficult to see. There would be a danger of electric shock. [2]
  c The water on the cable switch could cause an electric shock. [2]

Exam-style questions

Answers available at: www.hoddereducation.co.uk/cambridgeextras
22 Chose the correct fuse from 1 A, 3 A, 5 A, 13 A or 30 A for each of the following appliances. Take mains to be 220 V a.c. and remember $P = IV$.
  a 500 W microwave [2]
  b 2.5 kW heater [2]
  c 1.1 kW kettle [2]

Revision activity

Summarise this section into five key points about electrical safety.

4.5 Electromagnetic effects

4.5.1 Electromagnetic induction

Key objectives

By the end of this section, you should be able to:
- understand that a changing magnetic field linked to a conductor or a conductor moving in a magnetic field can induce an e.m.f. in the conductor
- describe an experiment which demonstrates electromagnetic induction and state the changes needed to increase the induced e.m.f.

- understand the e.m.f. is induced in a direction which opposes the change causing it
- use Fleming's right-hand rule to work out the relative directions of the force, magnetic field and induced current

When the magnetic field through a conductor changes, an e.m.f is induced. This is called **electromagnetic induction**. This change can be caused by:

- a conductor moving through a magnetic field (Figure 4.24)
- a magnetic field moving relative to a conductor (Figure 4.25)

Skills

Demonstrating electromagnetic induction

You should be able to describe an experiment to demonstrate electromagnetic induction.

When an e.m.f. is induced in a conductor which is part of a complete circuit, there is a current. You can measure the current using a sensitive centre-zero meter. A centre-zero meter is used because the current can be induced in either direction.

Demonstrating moving conductor: In Figure 4.24, an e.m.f. is induced only when the wire moves upwards (direction 1) or downwards (direction 2). The meter deflects in opposite directions in these two cases, but only when the wire is in motion. When moved in the other directions, the wire does not cut the magnetic field, so no e.m.f. is induced.

Demonstrating moving magnetic field: In Figure 4.25, an e.m.f. is induced only when the magnet is moving. If the magnet is pushed in, the meter deflects one way; if the magnet is pulled out, the meter deflects in the opposite direction. You can also change the direction of the induced current by reversing the poles of the magnet.

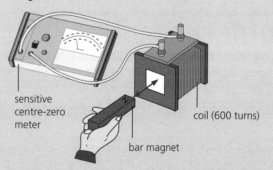

▲ **Figure 4.25 A voltage is induced in the coil when the magnet is moved in or out**

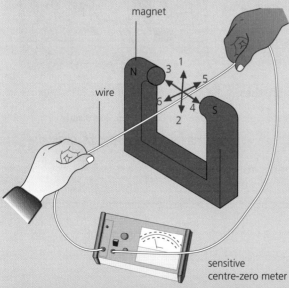

▲ **Figure 4.24 A voltage is induced when the wire is moved up or down in the magnetic field**

Remember an e.m.f. is always induced when the magnetic field through a conductor changes. If there is a complete circuit, a current will also be induced.

Factors affecting the size of an induced e.m.f.

The induced e.m.f. increases with an increase in:

- speed of relative motion of the magnet or coil
- number of turns of any coil
- strength of the magnet

The direction of the induced e.m.f. *opposes* the change that caused it.

In Figure 4.26, the moving magnet induces an e.m.f. in the coil, which causes a current in the coil. The current produces its own magnetic field, which opposes the movement of the magnet.

When the magnet moves down in Figure 4.26a, the top of the coil becomes a N pole. This repels the N pole of the magnet. When the magnet moves up in Figure 4.26b, the top of the coil becomes a S pole. This is to attract the N pole of the magnet

You should be able to state and use the relative directions of force, field and induced current. For a straight wire moving at right angles to a magnetic field, you use Fleming's right-hand (generator) rule.

Remember magnetic field direction is the direction of force felt by a N pole so the lines are in the direction from N to S.

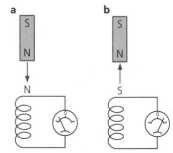

▲ Figure 4.26 The induced current opposes the motion of the magnet

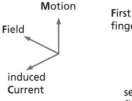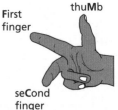

▲ Figure 4.27 The right-hand generator rule

4.5.2 The a.c. generator

Key objectives

By the end of this section, you should be able to:
- describe the structure of an a.c generator
- sketch and interpret graphs showing how the e.m.f. varies with time and relate this to the position of the coil

Figure 4.28 shows a simple a.c. generator. It is made up of a coil which can rotate between the poles of a magnet. The coil is connected to two slip rings which rotate with the coil. These are connected to the circuit via two fixed carbon brushes.

An e.m.f. is induced in the coil as it turns in the magnetic field. The wires on each side of the coil cut the field alternately moving up
and down, so the e.m.f. is induced in alternating directions. You can see how the e.m.f. varies with time in Figure 4.29.

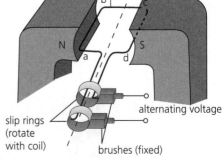

▲ Figure 4.28 A simple a.c. generator

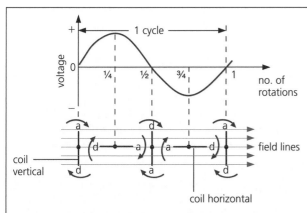

▲ **Figure 4.29 Output from a simple a.c. generator**

The peak e.m.f. occurs when the coil is horizontal, and the e.m.f. is zero when the coil is vertical. This is because at this point the coil is moving parallel to the magnetic field so no e.m.f. is induced. To understand why the induced e.m.f. changes direction, consider one side of the coil (ab). Initially ab is moving upwards and so induced current is in one direction. As the coil become vertical, side ab starts to move downwards and so the induced current is now in the opposite direction.

Sample question

10 Figure 4.30 shows the output from a simple a.c. generator.
  a Identify a point on the graph where the induced e.m.f. changes direction. [1]
  b Identify a point where the coil is perpendicular to the magnetic field (vertical). [1]
  c Describe what you would see if the coil rotated faster [2]

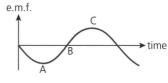

▲ **Figure 4.30**

Student's answers

a	B	[1]
b	A	[O]
c	The peak e.m.f. would be higher.	[1]

Teacher's comments

  a The student has correctly identified where the induced e.m.f. changes direction as the e.m.f. goes from negative to positive.
  b When the coil is vertical, the induced e.m.f. is zero. This is because the coil is moving parallel to the magnetic field.
  c The student realised the peak e.m.f. would be greater but they did not realise the frequency would increase. If there are two marks for a question, try to make two points.

Correct answers

a	B	[1]
b	B	[1]
c	The peak e.m.f. would be higher; the frequency would be greater.	[2]

Revision activity

Draw a labelled diagram of an a.c. generator and sketch the graph of how e.m.f. varies with time. Link the two diagrams to show the position of the coil at the peaks and troughs of the e.m.f.

4.5.3 Magnetic effect of a current

Key objectives

By the end of this section, you should be able to:
- describe the pattern of the magnetic field around a straight current-carrying wire and around a solenoid and describe an experiment to show these patterns
- describe how the magnetic effect of a current is used in loudspeakers and relays

- describe the variation in the magnetic field strength around wires and solenoids and how changing the magnitude and direction of the current affects their magnetic fields

Field due to a straight wire

A wire or coil carrying an electric current produces a magnetic field. The magnetic field pattern is a series of concentric circles, as shown in Figure 4.31. You can determine the direction of the magnetic field using the right-hand screw rule. The magnetic field lines point in the direction of rotation of the screw.

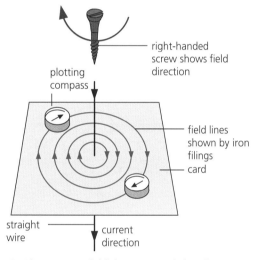

▲ **Figure 4.31 Field due to a straight wire**

Field due to a solenoid

A **solenoid** is a long cylindrical coil. When a current flows, the field pattern outside the solenoid is similar to that of a bar magnet. Inside the solenoid, there is a strong field parallel to the axis (Figure 4.32a). The right-hand grip rule gives the direction of the field.

The fingers of the right hand grip the solenoid pointing in the direction of the current and the thumb points to the N pole (Figure 4.32b).

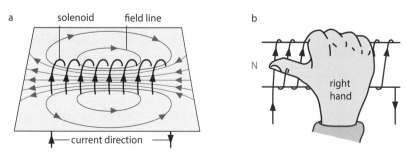

▲ **Figure 4.32 a) Field due to a solenoid and b) the right-hand grip rule**

Skills

Plotting the magnetic field around a current-carrying wire and solenoid

To observe the magnetic field patterns of a straight current-carrying wire or a solenoid, use iron filings and a plotting compass.

Magnetic field around a wire: Thread the wire through a piece of card held horizontally by a clamp stand (Figure 4.31). Sprinkle iron filings onto the card and then tap. The iron filings will show the shape of the field. You can use a plotting compass to determine the direction of the field.

Magnetic field around a solenoid: Thread a wire through a piece of card held horizontally (as shown in Figure 4.32a). Sprinkle iron filings onto the card and then tap. The iron filings will show the shape of the field. You can use a plotting compass to determine the direction of the field.

Variation of magnetic field strength

The magnetic field strength around a current-carrying wire is not constant. The magnetic field strength decreases with distance. You can see this in the magnetic field lines in Figure 4.31, which become further apart as you move away from the wire.

The closeness of the magnetic field lines in Figure 4.32a indicates the strength of the field within the solenoid. Outside the solenoid, the further away a point is, the weaker the field.

For both the straight wire and solenoid, the higher the current, the stronger the magnetic field. If the current reverses, the direction of the magnetic field is also reversed. You can also increase the strength of the solenoid's field by using more coils.

Applications of the magnetic effect of a current

A solenoid wrapped around an iron core forms an electromagnet. Electromagnets are used in cranes to lift iron objects and scrap iron, as well as in many electrical devices.

Relay

A relay is a device that enables one electric circuit to control another. It is often used when the first circuit carries only a small current (e.g. in an electronic circuit) and the second circuit requires a much higher current.

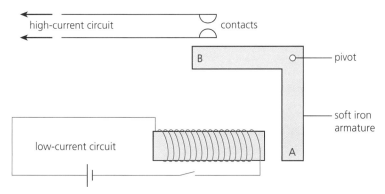

▲ **Figure 4.33 Magnetic relay**

When the switch is closed in the low-current circuit in Figure 4.33, current flows to the electromagnet, which attracts end A of the soft iron armature. The armature pivots and end B moves up to close the contacts in the high-current circuit. This circuit is now complete and the high current flows through the device, e.g. a motor, a heater or an alarm bell.

Loudspeaker

Figure 4.34 shows a loudspeaker. It consists of a circular permanent magnet with a central pole and a ring pole. A coil of wire sits over the ring pole and is attached to a paper cone.

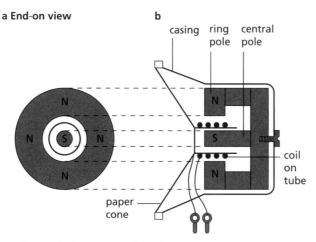

▲ **Figure 4.34 Moving-coil loudspeaker**

There is an alternating current in the coil. The changing magnetic field in this coil causes it to move up and down, making the cone vibrate and produce a sound. As the frequency of the alternating current changes so does the frequency of vibration, producing different frequency sound waves.

Sample question

11 Figure 4.35 shows an electric bell.

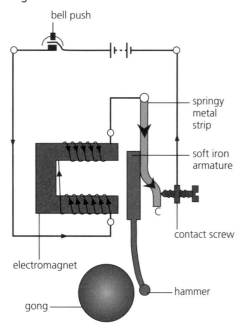

▲ **Figure 4.35**

For the electric bell shown in Figure 4.35:

a Describe what happens when the bell push is pressed. [4]
b Explain why iron is used for the armature. [1]
c Choose a suitable material for the core of the electromagnet. Give your reasons. [2]

Student's answers

a   When the bell is pressed, the electromagnet switches on. The iron
    armature is attracted and the hammer hits the gong.   [2]
b   A strong magnet is required.   [0]
c   Use soft iron, as the electromagnet must be switched on and off
    repeatedly.   [2]

Correct answers

a   When the bell push is pressed, a current flows through the
    electromagnet, which becomes magnetised. The armature is attracted
    to the electromagnet and the hammer strikes the gong. The
    movement of the armature breaks the circuit that applies current to
    the electromagnet. The armature is released and springs back. The
    circuit is re-made, the process repeats and the bell rings
    continually for as long as the bell push is pressed.   [4]
b   Iron is used because the armature must be attracted to the
    electromagnet.   [1]
c   Use soft iron, as the electromagnet must be switched on and off
    repeatedly.   [2]

Teacher's comments

a   The student has started the description well. However, they have not seen that when the hammer strikes the gong the circuit is broken. This switches off the electromagnet. The armature springs back remaking the circuit.
b   The armature needs to be made from a magnetic material.
c   Correct answer.

Revision activity

Create a revision poster on magnetic fields caused by currents – include how they are detected and how they can be used.

4.5.4 Force on a current-carrying conductor

Key objectives

By the end of this section, you should be able to:
● describe an experiment which shows how the force acting on a current-carrying wire in a magnetic field is affected by reversing the current or direction of the magnetic field

● use Fleming's left-hand rule to determine the directions of force, current and magnetic field relative to each other
● determine the direction of force on a beam of charged particles in a field

A wire or conductor carrying a current in a magnetic field experiences a force. The direction of the force depends on the direction of the magnetic field and the direction of the current.

Skills

Demonstrating force on a current-carrying wire
You can demonstrate the force on a current-carrying wire using the apparatus shown in Figure 4.36.

The wire is only loosely suspended and moves up when the current is switched on. The wire will move down if either the current is reversed or the magnet poles are swapped to reverse the field. If both the field and current are reversed, the wire will again move up.

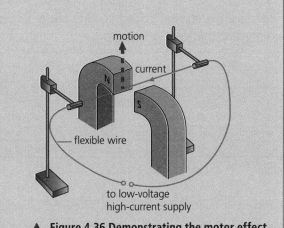

▲ **Figure 4.36 Demonstrating the motor effect**

Fleming's left-hand rule

You must be able to state and use Fleming's left-hand (motor) rule to determine the relative directions of force, magnetic field and current.

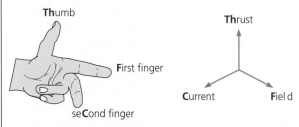

▲ Figure 4.37 Fleming's left-hand rule

Force on beams of charged particles in a magnetic field

A beam of charged particles experiences a force in a magnetic field. Figure 4.38 shows the path of an electron beam in a uniform magnetic field. As the force acts at right angles to the beam, it follows a circular path.

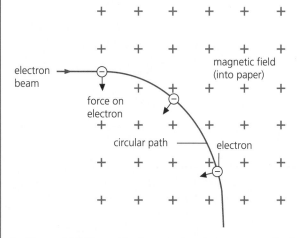

▲ Figure 4.38 The path of an electron beam perpendicular to a magnetic field

To determine the direction of the force, use Fleming's left-hand rule. The crosses show that the field is perpendicular to the paper and directed into it. They represent the back of an arrow. The important thing to remember is current direction is the direction of positive charge so point your second finger in the opposite direction to that of the electron beam – electrons are *negatively* charged.

To show the magnetic field coming out of the paper, use dots to represent the front of an arrow.

Sample questions

12 Figure 4.39 shows a wire in a magnetic field. The current through the wire is switched on.

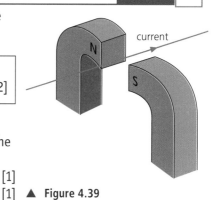

a State and explain the direction of the force on the wire when the current is switched on. [2]

b For each of the following changes, made one at a time, state whether the magnitude of the force on the wire increases, stays the same, decreases or decreases to zero:
  i current changes direction [1]
  ii current drops to zero [1]
  iii current increases [1]
  iv magnetic field increases [1]
  v magnetic field changes direction [1]

▲ **Figure 4.39**

Student's answers

a	Force is up, as in Figure 4.36.	[1]

b	i	Changes direction.	[0]
	ii	Becomes zero.	[1]
	iii	Increases.	[1]
	iv	Increases.	[1]
	v	Changes direction.	[0]

Teacher's comments

a The student did not observe that the current is reversed from Figure 4.36.

b The student seems in places to have committed the classic error of answering the question that was expected not what was asked.
  i The student answered a different question.
  ii Correct answer.
  iii Correct answer.
  iv Correct answer.
  v The student answered a different question.

Correct answers

a	Force is down by Fleming's left-hand rule.	[2]

b	i	Force stays the same.	[1]
	ii	Force decreases to zero.	[1]
	iii	Force increases.	[1]
	iv	Force increases.	[1]
	v	Force stays the same.	[1]

13 Figure 4.40 shows a beam of positively charged particles entering a magnetic field at right angles to it. Sketch the path of the beam and explain your answer. [3]

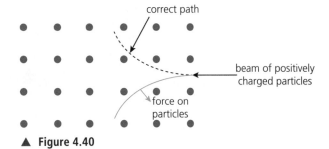

▲ **Figure 4.40**

Student's answer

The student's answer is shown by the blue line in Figure 4.40. The magnetic field is out of the page. The positively charged particles feel a force at right angles to their direction [1] so curve downward [0]. Using the left-hand rule [1 mark for diagram ignoring direction].

Teacher's comments

The student's statement about the magnetic field is correct, and they understand that there is a force at right angles to the direction of the beam. The blue line shows a curved path. The beam is positively charged and so direction of the current is to the left. The student knew they had to apply the left-hand rule to their directions but made a mistake.

Correct answer

The magnetic field is out of the page. The current is to the left, so according to Fleming's left-hand rule [1] the charges feel a force upwards [1] as they enter the magnetic field. The force is always perpendicular to the motion, so the path is curved as shown in Figure 4.40 by the black dotted line. [1]

4.5.5 The d.c. motor

Key objectives

By the end of this section, you should be able to:
- state that a current-carrying coil can experience a turning effect in a magnetic field and know the factors that increase the turning effect
- describe the structure of an electric motor and how it works

Turning effect on a coil

A straight wire in a magnetic field feels a force. If you make the wire into a rectangular coil and place it in a magnetic field, one side feels a force upwards as the other feels a force downwards. This causes a turning effect. Figure 4.41 shows a single coil in a magnetic field.

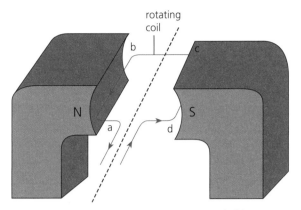

▲ **Figure 4.41 A single coil in a magnetic field**

You can see the current flows in opposite directions either side of the coil. The turning effect is increased by:

- increasing the number of turns on the coil
- increasing the current
- increasing the strength of the magnetic field

Simple d.c. electric motor

The directions of the forces on the coil are worked out using Fleming's left-hand (motor) rule. In Figure 4.42 there is an upwards force from the magnetic field on the wire ab. The current in the wire cd is in the other direction, so it experiences a force in the opposite direction (downwards). These two forces cause the coil of wire to turn clockwise.

The commutator and brushes act as a switching mechanism that changes the direction of the current every half turn to allow continuous rotation. When the coil is vertical the gaps in the split ring commutator line up with the brushes and there is no longer a current in the coil. The coil keeps moving as there is no force acting to stop it. Wire ab is now on the right, so moves down. Similarly, wire cd is now on the left and moves up.

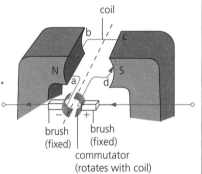

▲ **Figure 4.42 The commutator and brushes of a d.c. motor**

4.5.6 The transformer

Key objectives

By the end of this section, you should be able to:
- describe the construction of a simple transformer and explain how it works
- recall and use the transformer equation and use the terms *primary* and *secondary*, *step up* and *step down*

  - recall and use the equation for 100% efficiency in a transformer

- describe how transformers are used for high-voltage transmission and state their advantages

  - recall and use the correct power loss equation to explain why power losses are less at higher voltages

A **transformer** transforms (changes) an alternating voltage from one value to another of greater or smaller value. It consists of two coils wound on the same soft iron core. The primary coil is supplied with an alternating current and the secondary coil provides an alternating current to another circuit.

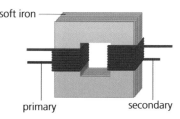

▲ **Figure 4.43 Primary and secondary coils of a transformer**

The a.c. in the primary coil sets up a changing magnetic field in the soft iron core. As the changing magnetic field cuts through the secondary coil, it induces an alternating e.m.f. in the secondary coil.

Transformers will only work with a.c. This is because if d.c. is used the current remains constant and so does the magnetic field. You need a changing magnetic field to induce an e.m.f.

Transformer equation

A step-up transformer has more turns on the secondary coil than the primary coil and the V_s is greater than V_p. In a step-down transformer, there are fewer turns on the secondary than the primary coil and V_s is less than V_p.

The relationship between the number of coils on the primary (N_p) and the number of coils on the secondary (N_s) is given by the equation:

$$\frac{V_p}{V_s} = \frac{N_p}{N_s}$$

When the p.d. is increased in a step-up transformer, the current is decreased by the same proportion. In an ideal transformer which has 100 % efficiency:

power in the primary = power in the secondary

$$I_p V_p = I_s V_s$$

As you can see, if the p.d. is halved then the current is doubled.

Transmission of electrical power

Electricity is transmitted over large distances at very high voltages, in order to reduce the energy losses due to the resistance of the transmission lines. This is achieved by having a step-up transformer at the power station to increase the voltage to several hundred thousand volts. Where the electricity is to be used, there is a series of step-down transformers to reduce the voltage to values suitable for use in factories or homes.

The advantages of high voltage transmission are:

- less power loss in the cables as the heating effect in the cables is less
- lower current in the cables means that thinner/cheaper cables can be used

Power loss in transmission cables

All power cables have some resistance. This means some energy is transferred to thermal energy as it is transmitted. The power loss in a cable (P) is given by the equation:

$$P = I^2 R$$

When the voltage is stepped up the current is stepped down. Power is therefore transmitted at the highest possible voltage in order to reduce the current and thus the losses in the cables.

Sample question

14 A transformer used by students in a school laboratory has 5500 turns on the primary coil and is supplied with 110 V a.c. The secondary coil has 500 turns.

    a Calculate the output voltage.   [3]

    b Explain the principle of operation of a transformer.   [3]

Student's answers

a $\dfrac{V_p}{V_s} = \dfrac{N_p}{N_s}$

$\dfrac{110}{V_s} = \dfrac{500}{5500}$

$V_s = \dfrac{5500 \times 110}{500} = 1210\,V$ [1]

b   The primary coil acts as an electromagnet supplied with an alternating current. This flows in the soft iron core. Therefore, the secondary coil has an alternating current.   [1]

Teacher's comments

a The student started with the correct equation, but muddled up the primary and secondary coil when they substituted the numbers into the equation. The student should have realised that students in a laboratory would never have access to such a high voltage, so something must have gone wrong in the calculation.

b The student knows that there is a magnetic field but completely fails to use this information. The answer also gives the impression that current flows through the core to the secondary coil, which is completely wrong.

Correct answers

a $\dfrac{V_p}{V_s} = \dfrac{N_p}{N_s}$

$\dfrac{110}{V_s} = \dfrac{5500}{500}$

$V_s = \dfrac{500 \times 110}{5500} = 10\,V$ [3]

b   The alternating current in the primary coil produces an alternating magnetic field in the soft iron core. The secondary coil is in this alternating magnetic field, so an alternating e.m.f. is induced.   [3]

Exam-style questions

Answers available at: www.hoddereducation.co.uk/cambridgeextras

23 The wire in Figure 4.24 is moved upwards.
 a Describe how to increase the e.m.f. induced in the wire. [3]

 b State whether the induced current is in a clockwise or anticlockwise direction in the wire. [1]
24 The magnet in Figure 4.44 is released and falls away from the coil. The needle on the centre-zero meter moves to the left.
 Describe what would happen to the needle on the centre-zero meter if:
 a the S pole of the magnet is moved upwards [1]
 b the N pole of the magnet is moved upwards [1]
 c the coil is moved down over the magnet as it is shown in Figure 4.44 [1]

▲ **Figure 4.44**

25 A loudspeaker is made up essentially of a stationary magnet that is close to a small coil fixed to a paper cone. The signal from the amplifier is a small alternating current supplied to the coil. Describe briefly:
 a the variation of the magnetic field produced by the coil [1]
 b the variation of the magnetic force on the coil [1]
 c the motion of the paper cone [1]
26 Describe an experiment to show that a force acts on a current-carrying wire in a magnetic field and what happens if the current is reversed. [3]
27 Figure 4.45 shows a coil with several turns carrying a current in a magnetic field.
 a State the effect that the current has on the coil. [1]
 b State whether the size of this effect is increased, the same, decreased or decreased to zero when:
  i the current is reversed
  ii the current is increased
  iii the magnets are removed
  iv the number of turns is increased [4]

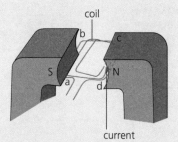

coil

current

▲ **Figure 4.45**

28 Figure 4.46 shows a coil that can rotate in a magnetic field.
 a State the direction of any forces on:
  i wire ab
  ii wire cd [2]
 b The coil is rotated so that it is vertical. State the direction of any forces now acting on:
  i wire ab
  ii wire cd
  Explain how you reached your answers. [4]

Revision activity

Create a spider diagram about the transformer. Write down these words and find links between them: transformer, step up, step down, high-voltage transmission, power losses.

c The coil is rotated so that it is again horizontal with wire ab on the right and wire cd on the left. State the direction of any forces now acting on:
  i wire ab
  ii wire cd [2]

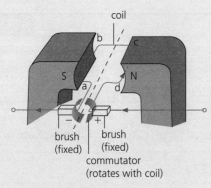

coil

S   N

brush (fixed)   brush (fixed)
commutator (rotates with coil)

▲ **Figure 4.46**

29 A transformer is used to provide an a.c. 6V supply for a laboratory from 240V a.c. mains. The secondary coil of the transformer has 100 turns. Calculate the number of turns on the primary coil. [3]

30 A transformer has 1200 turns on the primary and 20 turns on the secondary coil. The input voltage is 120V a.c and the current in the primary coil is 10mA. Calculate the current in the secondary coil. [4]

31 Transmission cables have a resistance of 400Ω. Calculate the power loss in the cables:
  a when the current is 2.5A [2]
  b when the current is 250A [2]

5 Nuclear physics

Key terms

Term	Definition
Alpha-particle (α)	Radiation consisting of helium ions with a double positive charge (^4_2He)
Atom	Tiny constituent of matter
Background radiation	Ever-present radiation resulting from cosmic rays from outer space and radioactive materials in rocks, the air, buildings
Beta-particle (β)	Radiation consisting of high-speed electrons ($^0_{-1}\text{e}$)
Electron	Negatively charged elementary particle ($^0_{-1}\text{e}$)
Gamma-radiation (γ)	High-frequency, very penetrating electromagnetic waves
Half-life	The average time for half the nuclei in a radioactive sample to decay
Ion	Charged atom or molecule that has lost or gained one or more electrons so that it is no longer neutral
Isotope	One form of an element that has the same number of protons but a different number of neutrons in the nucleus from other isotopes of the same element
Neutron	An uncharged subatomic particle found in the nucleus of an atom (except that of hydrogen)
Nucleon number, A	Number of protons and neutrons in the nucleus
Proton	Positively charged particle found in the nucleus of an atom
Proton number, Z	Number of protons in the nucleus
Fission	The break-up of a large nucleus into smaller parts
Fusion	The union of light nuclei into a heavier one

5.1 The nuclear model of the atom

5.1.1 The atom

Key objectives

By the end of this section, you should be able to:
- describe the structure of an atom
- state how positive and negative ions are formed

- describe how the alpha scattering experiment supports the nuclear model of the atom

The atom

The **atom** is the smallest particle of an element. It is made up of a central nucleus, with all the positive charge and nearly all the mass, and negatively charged electrons in orbit. The nucleus is very much smaller than the electron orbits, so the majority of every atom is empty space.

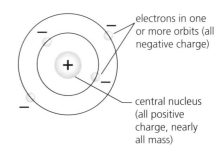

electrons in one or more orbits (all negative charge)

central nucleus (all positive charge, nearly all mass)

▲ Figure 5.1 The nuclear atom

The model of a nuclear atom was confirmed by observing a beam of α-particles (positively charged particles) travelling towards a sheet of thin metal foil.

● The vast majority of α-particles passed straight through without being deflected. This is evidence that most of the atom is empty space and the nucleus is very small.

● A few α-particles were deflected, some through a large angle, and a very small proportion bounced back (see Figure 5.2). This is evidence that the nucleus is positively charged because the positively charged alpha particles were strongly repelled. The small number deflected shows that all the mass and positive charge is concentrated in a small part of the atom – the nucleus.

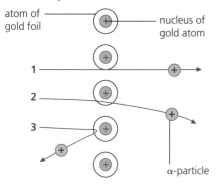

▲ **Figure 5.2 Scattering of α-particles by thin gold foil**

Path 1 is a long way from any nucleus and the α-particle is undeflected.

Path 2 is close to a nucleus and there is some deflection.

Path 3 heads almost straight for a nucleus and the α-particle rebounds back.

Ions

Atoms are neutral. They contain an equal number of positively charged **protons** and negatively charged **electrons**. An atom with a charge is called an **ion**. If an atom gains electrons, it becomes negatively charged and is called a negative ion. If an atom loses electrons, it becomes positively charged and is called a positive ion.

5.1.2 The nucleus

Key objectives

By the end of this section, you should be able to:
● define proton number Z and nucleon number A and calculate the number of neutrons in the atom
● use nuclide notation
● explain what is meant by an isotope

● describe the processes of nuclear fission and fusion including the nuclide equations
● relate the relative charge and relative mass of a nucleus to the proton and nucleon number, respectively

Protons and neutrons

A nucleus contains **protons** and **neutrons** which are known as the **nucleons**. To compare the particles inside an atom, you consider their mass and charge relative to each other. Table 5.1 shows relative mass, relative charge and their position.

▼ **Table 5.1 The relative mass, charge and position of the subatomic particles**

Particle	Relative mass	Relative charge	Position
Proton	Approximately 2000	+1	Inside the nucleus
Neutron	Approximately 2000	0	Inside the nucleus
Electron	1	−1	Outside the nucleus

Figure 5.3 shows a model of how the particles are arranged.

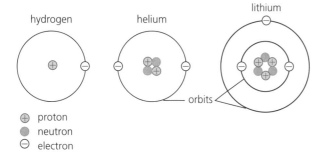

▲ **Figure 5.3 Protons, neutrons and electrons in atoms**

The number of protons in a nucleus is called the **proton number (Z)** and is the same as the number of electrons in orbit.

The number of nucleons (protons and neutrons) is called the **nucleon number (A)**. The difference between the nucleon number A and the proton number Z gives the number of neutrons in the nucleus ($A - Z$).

The nuclide (type of nucleus) of an element can be written with the notation $^A_Z X$, where X is the chemical symbol for the element.

Mass and charge on a nucleus

Different atoms have different masses. The **relative mass** of a nucleus depends on the number of nucleons. You use the nucleon number, A, to determine the relative mass. For example, oxygen, $^{16}_8 O$ has a relative mass of 16.

The **relative charge** of a nucleus is always positive and depends on the number of protons in the nucleus. You use the proton number Z to determine the relative charge. For example, an oxygen nucleus has a relative charge of +8.

Skills

Determining the number of neutrons in a nucleus

Aluminium has the symbol Al and is written as $^{27}_{13} Al$. Calculate the number of neutrons in the nucleus of an aluminium atom.

$A = 27$, $Z = 13$

number of neutrons = $A - Z = 27 - 13 =$ 14 neutrons

Isotopes and nuclides

Isotopes of the same element are different forms that have the same number of protons but different numbers of neutrons in the nucleus. An element may have more than one naturally occurring isotope.

Some radioactive isotopes occur naturally, e.g. carbon-14 is produced in the atmosphere by cosmic rays.

Many radioactive isotopes are produced artificially in nuclear reactors and have a wide range of practical uses, e.g. as a source of radiation to kill cancers (see p. 144) or as tracers in the human body or in a pipeline.

Energy from nuclear reactions

In **fission**, a heavy nucleus is split into smaller nuclei and some neutrons. In **fusion**, smaller nuclei join together to make a larger nucleus. In both processes, the mass of the starting atoms is greater than the products. The missing mass or mass defect is converted into energy.

Nuclear fission

In fission, a neutron strikes a large nucleus and it splits into two smaller nuclei, approximately the same size, and two or three more neutrons, for example,

$$^{235}_{92}\text{U} + {}^{1}_{0}\text{n} \longrightarrow {}^{144}_{56}\text{Ba} + {}^{90}_{36}\text{Kr} + 2{}^{1}_{0}\text{n}$$

▼ **Table 5.2 Nucleon and proton numbers of each nuclei in the fission reaction**

A	235	1	= 236	144	90	2	= 236
Z	92	0	= 92	56	36	0	= 92

Notice that the total values of A and Z on both sides of the equation are equal. As more neutrons are released in the reaction, these can go on to fission other uranium nuclei and start a chain reaction (Figure 5.4).

In an atomic bomb, the chain reaction is uncontrolled and leads to an explosion. In a nuclear reactor, the number of neutrons is carefully controlled. The lighter nuclei produced are themselves highly radioactive nuclear waste, which is difficult and expensive to dispose of.

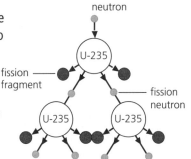

▲ **Figure 5.4 Chain reaction**

Nuclear fusion

Fusion occurs under conditions of extremely high temperature and pressure when light nuclei can join together. The nuclei need enough kinetic energy to overcome their electrostatic repulsion. Remember nuclei are positively charged and so will repel strongly. Fusion releases a large amount of energy.

The following reaction occurs in the Sun and other stars, as well as in the hydrogen bomb:

$$^{1}_{1}\text{H} + {}^{2}_{1}\text{H} \rightarrow {}^{3}_{2}\text{He}$$

Research reactors are currently experimenting into ways of maintaining controlled fusion reactions for possible power stations of the future

Sample question

1 The most common isotope of carbon is carbon-12, written ${}^{12}_{6}\text{C}$ in nuclide notation.

  a Write down the nucleon and proton numbers of carbon-12. [2]

  b Write down the number of electrons in a neutral atom of carbon-12. [1]

  c Carbon-14 is a radioactive isotope that exists in small quantities in the atmosphere. Write down the nucleon and proton numbers of carbon-14. [1]

  d Write down the nuclide notation for carbon-14. [2]

  e Work out the number of neutrons in a nucleus of carbon-14 and state the difference between the nuclei of carbon-12 and carbon-14. [2]

Student's answers

> a nucleon number, A = 12
>   proton number, Z = 6   [2]
> b number of electrons = 12   [0]
> c nucleon number, A = 14
>   proton number, Z = 6   [1]
> d $^{6}_{14}C$ [0]
> e A carbon-14 nucleus is bigger, with more particles.   [0]

Teacher's comments

a Correct answers.
b The student has incorrectly thought that the number of electrons is the same as the number of nucleons.
c Correct answer.
d The student has mixed up the nucleon and proton numbers – care needs to be taken here!
e The first part of the question has not been answered and the rest of the answer is too vague.

Correct answers

> a nucleon number, A = 12
>   proton number, Z = 6   [2]
> b number of electrons = number of protons = 6   [1]
> c nucleon number, A = 14
>   proton number, Z = 6   [1]
> d $^{14}_{6}C$
> e number of neutrons in carbon-14 nucleus
>   = nucleon number – proton number   [2]
>   = A – Z = 14 – 6 = 8
>   number of neutrons in carbon-12 nucleus = 12 – 6 = 6
>   A carbon-14 nucleus has two extra neutrons.   [2]

Exam-style questions

Answers available at: www.hoddereducation.co.uk/cambridgeextras

1 Copy and complete the table to indicate the composition of an atom of each of the isotopes of strontium given.   [6]

Isotope	Number of protons	Number of neutrons	Number of electrons
$^{88}_{38}Sr$			
$^{90}_{38}Sr$			

2 Sodium (symbol Na) has 11 protons and 12 neutrons.
   a Write down the nuclide notation for sodium.   [3]
   b A sodium atom loses one electron. State the charge of the ion formed.   [1]
   c Write the relative mass and relative charge of a sodium nucleus.   [2]
3 a Explain the difference between nuclear fission and nuclear fusion.   [4]
   b State one significant similarity between them.   [1]
   c State where a fission reaction occurs.   [1]
   d State where a fusion reaction occurs.   [1]
4 Copy the nuclear equation for the fission of uranium-236 and fill in the gaps to balance the equation.   [2]

$$^{236}_{92}U + ^{1}_{0}n \rightarrow ^{144}_{\underline{}}Ba + ^{\underline{}}_{36}Kr + 3^{1}_{0}n$$

Revision activity

Create a poster showing the structure of an atom and how this is represented using nuclide notation. Include descriptions of how positive and negative ions are formed and what an isotope is.

Create a table to match the descriptions of a nuclear model to the evidence from the alpha scattering experiment. Describe how energy is released in nuclear fission and fusion.

5.2 Radioactivity

5.2.1 Detection of radioactivity

Key objectives

By the end of this section, you should be able to:
- state what is meant by background radiation and the main sources
- state how a detector connected to a counter is used to measure ionising nuclear radiation and the units of count rate are counts/s or counts/minute

- determine the corrected count rate using measurements of background radiation

There is radiation all around you all of the time. This is called **background radiation**. This is mainly due to natural sources such as radon gas in the air, cosmic rays, rocks and buildings and from food and drink. The value varies depending on where you live.

In collisions between radioactive particles and molecules in the air, the radioactive particles knock electrons out of the atoms, leaving the molecules positively charged. This is called **ionisation**. This ionising effect is used to detect radiation.

In a Geiger–Müller tube, the ionising radiation causes a pulse of current to flow between the electrodes. The tube is connected to a counter which counts these pulses of current and gives the total in a set amount of time. This is used to calculate the count rate which is measured in counts per second or counts per minute.

If you are measuring the count rate for a radioactive source, you need to correct for background radiation. To do this you simply subtract the count rate due to background radiation from your reading. To measure the background count, take a reading using the detector for a few minutes when the radioactive source is not in the room. Divide the total count by the time to determine the background count rate.

5.2.2 The three types of nuclear emission

Key objectives

By the end of this section, you should be able to:
- describe the emission of radiation from a nucleus as spontaneous and random in direction
- identify alpha, beta and gamma emissions from their basic characteristics

- describe the deflection of α-particles, β-particles and γ-radiation in electric and magnetic fields
- explain the relative ionising effects of each type of emission

Alpha, beta and gamma radiation

Radioactivity occurs when an unstable nucleus decays and emits one or more of the three types of radiation: α (alpha)-particles, β (beta)-particles or γ (gamma)-radiation. Radioactivity is a random process. It is impossible to know when an individual radioactive nucleus will decay and in what direction. It is also a spontaneous process. This means it happens on its own and is not affected by factors such as temperature or pressure.

Table 5.3 shows how you can identify the different types of radiation from their properties

▼ Table 5.3 The properties of α, β and γ radiation

Emission	Nature	Charge	Penetration	Ionising effect
α-particle	Helium nucleus (two protons and two neutrons)	+2	Stopped by thick paper or a few centimetres of air	Very strong
β-particle	High-speed electron	–1	Stopped by a few millimetres of aluminium	Weak
γ-radiation	Electromagnetic radiation	None	Only stopped by many centimetres of lead	Very weak

Explaining range and ionising effect of alpha, beta and gamma

Alpha-particles have the largest mass and the highest charge. This means they are much more likely to interact with matter and so they cannot penetrate very far. In collisions, the positive charge easily knocks outer electrons from atoms and so they have a high ionising power. In each collision, the α-particle transfers energy from its kinetic energy store and so slows down and has a short range.

Beta-particles have less mass than α-particles and so are less likely to interact with matter. This means they can penetrate further. The single negative charge means they are able to knock out electrons from atoms as the particles pass them and so have ionising ability. As they make fewer collisions, they can travel a greater distance (a few metres through air) before all of their kinetic energy has been transferred through collision.

Gamma-radiation has no charge as it is electromagnetic radiation. It has very little interaction with matter. Gamma-radiation can still knock an outer electron from an atom and ionise it. However, it is the least likely to ionise. As it has the least interaction with matter, it travels a long distance before transferring all its energy.

Deflection of alpha, beta and gamma in magnetic and electric fields

In a magnetic field, alpha and beta are deflected according to their charge. You can use Fleming's left-hand rule to determine the direction. Remember, conventional current is in the direction of flow of positive charge. Gamma, as it is uncharged, is unaffected by the magnetic field. Their deflections in a magnetic field are summarised in Figure 5.5a. Use Fleming's left-hand rule to check the directions shown.

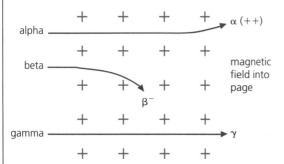

▲ Figure 5.5a Deflection of α- and β-particles and γ-rays in a magnetic field

In an electric field, α-particles are attracted towards the negative plate, β-particles towards the positive plate and γ-rays pass through with no deflections. The deflections are summarised in Figure 5.5b.

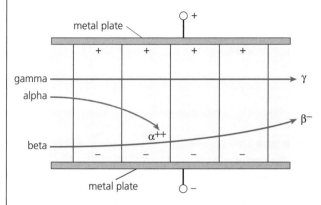

▲ **Figure 5.5b Deflection of α- and β-particles and γ-rays in an electric field**

5.2.3 Radioactive decay

Key objectives

By the end of this section, you should be able to:
- state that radioactive decay is a change to an unstable nucleus and can lead to emission of alpha, beta or gamma radiation and that it is both random and spontaneous
- know that during alpha and beta decay the nucleus changes to that of a different element

- understand that isotopes may be radioactive because of an excess of neutrons and/or because the nucleus is too heavy
- describe how the changes to a nucleus improve its stability and to use radioactive decay equations using nuclide notation

Radioactive decay is when an unstable nucleus emits radiation. When the nuclei emit α-particles or β-particles, the nuclei change to a different element. This element may also be unstable and emit radiation.

Alpha-decay
An α-particle is a helium nucleus. It consists of two protons and two neutrons. During α-decay, the nucleus loses two neutrons and two protons. The nucleon number goes down by four. The proton number goes down by two, so the nuclide changes to another element. An example of α-decay can be shown by a word equation:

radium-226 → radon-222 + α-particle

The same example of α-decay can be shown by an equation in nuclide notation:

$$^{226}_{88}\text{Ra} \rightarrow {}^{222}_{86}\text{Rn} + {}^{4}_{2}\text{He}$$

Note that, because an α-particle is the same as a helium nucleus, it is shown as He in nuclide notation.

Beta-decay

In β-decay, a neutron in the nucleus changes to a proton and an electron, which is emitted at high speed as a β-particle. The nucleon number is unchanged. The proton number goes up by one, so the nuclide also changes to another element.

An example of β-decay can be shown by word and nuclide equations:

carbon-14 → nitrogen-14 + β-particle

$$^{14}_{6}C \quad \rightarrow \quad ^{14}_{7}N \quad + \quad ^{0}_{-1}e$$

Note that, because a β-particle is an electron, it is shown as e in nuclide notation, with a nucleon number of 0, because it has negligible mass, and a proton number of –1, because of its negative charge.

Gamma-emission

Gamma-radiation is an electromagnetic wave and so does not change the structure of the nucleus. However, γ-radiation is usually given off during both α-decay and β-decay, and can be added to the equations, for example:

$$^{226}_{88}Ra \rightarrow {}^{222}_{86}Rn + {}^{4}_{2}He + \gamma$$

Skills

Writing a decay equation

Each nuclide equation of a radioactive decay balances. The sum of the nucleon number and proton number on the left-hand side of the equation is the same as on the right-hand side. You can use this to determine the nucleon and proton number of the nucleus after the decay.

For example, the nuclear equation shows the decay of an isotope of polonium. Calculate the values of A and Z.

$$^{210}_{84}Po \rightarrow {}^{A}_{Z}Pb + {}^{4}_{2}He$$

nucleon number: $210 = A + 4$

$$A = 206$$

proton number: $84 = Z + 2$

$$Z = 82$$

Nuclear stability

An isotope of an element may be unstable because it has too many neutrons or because it is too heavy (has too many nucleons). Figure 5.6 shows a graph of the number of neutrons (N) against the number of protons (Z). The stable nuclei lie on the stability line.

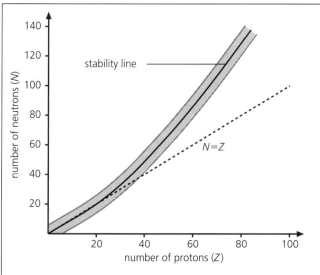

▲ **Figure 5.6 Stability of nuclei**

For unstable nuclides, a decay helps them move closer to the stability line. If the isotope lies above the line, then it needs to reduce the number of neutrons. It can do this through β-decay. Large nuclei often decay via α-emission to reduce the mass of their nucleus and become more stable.

Gamma-emission often happens after α- or β-decay as the nucleus is left in an excited state. Gamma-emission makes the nucleus more stable by releasing this excess of energy.

Sample question
REVISED

2 Radioactive strontium-90 (Sr, proton number 38) decays to yttrium (Y), emitting a β-particle and γ-radiation. Show this decay reaction as a nuclide equation. [4]

Student's answer

$$^{90}_{38}Sr \rightarrow {}^{89}_{39}Y + {}^{0}_{-1}e + \gamma \qquad [3]$$

Teacher's comments

The student has mostly got the answer right. The nuclide symbol for strontium is correct, as are the symbols for the β-particle and γ-radiation. The student has also correctly deduced that the proton number of yttrium is 39, one more than that of strontium. However, the nucleon number must stay the same in β-decay, as the nucleus has lost a neutron and gained a proton.

Correct answer

$$^{90}_{38}\text{Sr} \rightarrow {}^{90}_{39}\text{Y} + {}^{0}_{-1}\text{e} + \gamma \qquad [4]$$

Revision activity

Create flash cards with a question on one side and answer on the reverse. Questions should cover the sources of background radiation, the nature, properties and ionising effects of alpha, beta and gamma radiation, and the meaning of random and spontaneous. Use the cards to test yourself. Include decay equations for supplemental.

It is illegal to photocopy this page

5.2.4 Half-life

Key objectives

By the end of this section, you should be able to:
- define half-life and use this definition to perform half-life calculations including using data from table or decay curves

- perform half-life calculations taking into account background radiation
- explain how a radioactive isotope can be suitable for a particular use

Radioactive decay is a random process. It is impossible to predict when an individual nucleus will decay. However, *on average,* there is a definite decay rate for each isotope.

The decay rate is expressed as the **half-life**, which is the time for half the nuclei in a sample to decay. As it is hard to count the number of nuclei, you can determine the half-life by monitoring the disintegrations per second (activity) from a radioactive source. In Figure 5.7 the half-life is 10 minutes. This is because the disintegrations per second halves every 10 minutes.

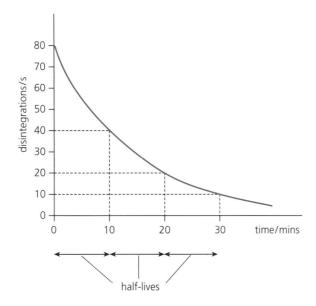

▲ **Figure 5.7 The half-life of a material can be found by using a graph (decay curve)**

Extended candidates need to be able to carry out half-life calculations allowing for background radiation.

Skills

Plotting a decay curve

You need to know how to plot and interpret decay curves to determine the half-life of an isotope.
- When you plot a decay curve, the time is always on the x-axis and the number of disintegrations per second or count rate is on the y-axis.
- Choose a scale so you can plot the points easily and so you use as much of the graph paper as possible.
- The line of best fit must be a smooth curve.
- You can then use the curve to find the half-life. Check the value by finding the time taken to halve the activity in more than one place.

> ● To correct for background radiation, subtract the background count rate from each value before you plot the curve.

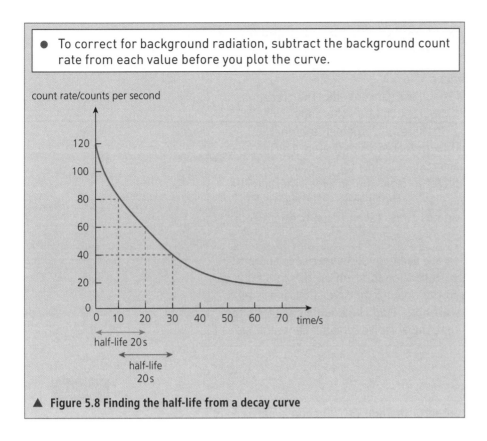

count rate/counts per second

▲ Figure 5.8 Finding the half-life from a decay curve

Uses of radioactivity

The type of radiation emitted and the half-life of the radioactive isotope determine how it can be used.

Smoke alarm

Inside a smoke alarm, there is a small radioactive source which emits α-particles. The α-particles ionise the air and a small current flows between electrodes. When there is smoke, the ions cannot flow freely between the electrodes. The current decreases and this is detected by a circuit, which sounds the alarm. An α-emitting source is chosen because α-particles are highly ionising and not very penetrating. This means the smoke alarm is not dangerous to anyone nearby. A source with a long half-life is used so that the activity remains constant.

Irradiation and sterilisation

Radiation causes ionisation and can kill living cells. This makes radiation useful for killing bacteria on food to make it last longer. This is called irradiation. The food is not radioactive as there is no radioactive material on the food. Radiation is also used to sterilise medical instruments. Gamma-radiation is used because it is the most penetrating so can pass through the packaging. A source with a long half-life is used so that the activity remains constant.

Thickness monitoring

A radioactive source is placed on one side of a strip of paper, plastic or metal during its manufacture. A radiation detector on the other side of the strip monitors how much has been absorbed. This indicates the thickness of the strip. Beta-emitters are used for thinner materials and γ-emitters for thicker materials as they can penetrate further and can still be detected on the other side. A source with a long half-life is used so that the activity remains constant.

Diagnosis and treatment of cancer

Radioactive tracers can be used to detect cancer cells. The tracer is injected into the body and its progress through the body can be detected by a gamma camera. If there is a problem, a higher concentration of γ-radiation will be detected in that area. The radioactive isotopes used are gamma as it is low ionising and can penetrate the body, so can be detected from the outside. The half-life will be short so that it will have a low activity soon after the test is over but long enough for the readings to be taken. Usually about four hours.

High-energy beams of γ-radiation are used to kill cancer cells. Careful arrangements are made to concentrate the radiation on the cancer cells and not kill other healthy cells in the body. Extensive shielding is also needed to protect the medical staff operating the equipment. A source with a long half-life is used so that the activity remains constant.

Sample question

REVISED

3 A radioactive sample gives a detector reading of 700 counts per second. The half-life of the sample is seven days.
   a Work out the expected detector reading two weeks later. [2]

   b The background count is 100 counts per second. Calculate the expected detector reading taking background into account. [2]

Student's answer

a After 1 week, detector reading $= \dfrac{700}{2} = 350\,counts/s$

After 2 weeks, detector reading $= \dfrac{350}{2} = 175\,counts/s$ [2]

b background count is 100 counts/s so detector reading
= 175 + 100 = 275 [0]

Teacher's comments

a The student has correctly calculated the count rate after two weeks.

b The student has not allowed for the background count correctly. You have to subtract background count from the initial reading. Determine the count rate after two weeks and then add background count on to the value.

It is illegal to photocopy this page

Correct answers

a After 1 week, detector reading = $\dfrac{700}{2}$ = 350 counts/s

After 2 weeks, detector reading = $\dfrac{350}{2}$ = 175 counts/s [2]

b Initial detector reading due to sample = 700 − 100 = 600 counts/s

After 1 week, detector reading due to sample = $\dfrac{600}{2}$ = 300 counts/s

After 2 weeks, detector reading due to sample = $\dfrac{300}{2}$ = 150 counts/s

Final detector reading including background = 150 + 100 = 250. [3]

Revision activity

Check you understand what is meant by half-life, and you can do simple half-life calculations by creating flash cards to test these out in your class and then switching with a partner.

Work in a group of three. Each of you choose alpha, beta or gamma and then prepare a 30 second presentation on a practical use of your radioactive source. The presentation must include an explanation of why that type of radiation is chosen and whether a long or short half-life is needed. Give the presentation to your group.

5.2.5 Safety precautions

Key objectives

By the end of this section, you should be able to:
- state the effects of ionising nuclear radiation on living things
- describe precautions for handling, using and storing radioactive materials safely and explain these precautions.

Radioactive emissions are ionising radiation. Ionising radiation can be very harmful to living things. It either kills cells outright or mutates cells, which can lead to cancers. To reduce the risk of harm from ionising radiation, a number of simple precautions can be taken.

The following safety precautions should therefore be taken:

- Whenever possible, radioactive samples are in sealed casings so that no radioactive material can escape.
- Samples are stored in lead-lined containers in locked storerooms.
- Samples are handled only by trained personnel and must always be supervised when not in store.
- Radioactive samples are shielded and kept at as great a distance as possible from people. In the laboratory, they are handled with long tongs and students are kept at a distance. In industry, they are usually handled by remote-controlled machines.
- Workers in industry are often protected by lead and concrete walls, and wear film badges that record the amount of radiation received.

The safety precautions are designed to reduce exposure to the ionising radiation by:

- reducing the time people spend near the sources
- keeping the distance from the source and the person as large as possible
- by using suitable shielding to absorb the radiation

Sample question

4 An extremely strong source of α-particles and γ-rays is used in an experiment being demonstrated to a group of student observers. The source is held and moved by a robot arm controlled by a technician who is always at least 1 m away from the source. The observers are always at least 3 m away from the source.

   a These precautions are insufficient for the technician and for the students. Explain this. [2]

   b Suggest practicable improvements that would permit the demonstration to continue and be observed in a safe way. [2]

Student's answers

   a The students would be safe at that distance but the technician needs to move further away. [0]

   b The technician could use a video camera. [1]

Correct answers

   a The robotic handling and distance from the source protects the technician and students observers from the α-radiation. [1] Because of the strong γ-radiation, even at 3 m distance, this is extremely unsafe for both the technician and the observers. [1]

   b The source should be shielded by thick lead or concrete from all humans. [1] The experiment could be viewed by video camera. A remote screen behind the shielding would allow the technician to control the robot and the students to observe in complete safety. [1]

Teacher's comments

   a This answer makes no distinction between α- and γ-radiation. There is no risk of harm from the alpha-source. But 3 m of air will not protect from the gamma-source.

   b The answer shows some awareness of the possibility of remote observation but does not address the issue of appropriate shielding.

Exam-style questions

Answers available at: www.hoddereducation.co.uk/cambridgeextras

5 State what is meant by background radiation and name two natural sources. [3]

6 A Geiger–Müller tube is used to measure the background radiation in the room before an experiment. The total count from the detector after 2 minutes is 24.

   a Calculate the count rate in counts/s due to background radiation. [2]

   b A radioactive source has a count rate of 40 count/s. Calculate the corrected count rate. [1]

7 The penetration of two radioactive samples is tested by measuring the count rate with various types of shielding between the sample and the counter. The numbers in the table below indicate the count rate (CR) with each type of shielding in place: no shielding, thick card, 3 mm of aluminium (Al), 20 cm of lead (Pb).

Revision activity

Create a simple poster explaining the dangers of ionising radiation and describing how to move, use and store radioactive sources in a safe way.

Explain how these reduce the risk.

Copy the table and tick the appropriate boxes in the right-hand three columns to show the type or types of emission from that sample.　[3]

Sample	CR (none)	CR (card)	CR (Al)	CR (Pb)	α	β	γ
1	6000	1000	1000	20			
2	3000	3000	20	20			

8　A positron is a subnuclear particle with the same mass as an electron but with a positive charge. A certain nuclear reaction emits positrons and γ-rays, which are directed to pass parallel to and between two horizontal plates in a vacuum. The upper plate has a very high positive potential relative to the lower plate. Describe the path between the plates of:

　a　the positrons　[1]

　b　the γ-radiation　[1]

9　Radioactive uranium-238 (U, proton number 92) decays to thorium (Th), emitting an α-particle and γ-radiation. Show this decay reaction as a nuclide equation.　[4]

10　Caesium $^{137}_{55}$Cs decays by an emission of a β-particle to an isotope of barium (Ba). Write down the nuclide equation for this decay.　[3]

11　Describe how the emission of a β-particle can increase the stability of a radioactive isotope.　[3]

12　The results in the table below came from an experiment to determine the half-life of a radioactive sample.

Time/min	0	2	4	6	8
Counts/s	400	280	200	140	100

　a　Use the data in the table to determine the half-life of the radioactive isotope.　[1]

　b　Calculate the fraction of the sample left after 12 minutes.　[1]

13　In an experiment to find the half-life of protactinium, a student gained the results given in the table below.

Time/s	0	30	60	90	120
Count/s	1200	895	670	504	377

　a　Plot a graph to show the decay curve for this data.　[4]

　b　Use your graph to determine the half-life of protactinium.　[2]

14　The table below shows the half-life and type of emissions from three radioactive sources A, B and C.

Radioactive isotope	Type of emission	Half-life
A	β-particles	25 years
B	γ-radiation	28 years
C	γ-radiation	3 hours

　a　Explain which source could be used as a medical tracer.　[3]

　b　Explain which source could be used to monitor the thickness of thin sheets of aluminium.　[3]

15　A teacher is using a radioactive source to demonstrate the absorption of γ-radiation in air. The source is shielded so radiation is only emitted from one side.

　a　Describe some precautions the teacher could take to keep themselves and the students safe.　[2]

　b　Explain how these precautions keep the teacher and students safe.　[2]

6 Space physics

Key terms

Term	Definition
Big Bang Theory	Violent explosion of the Universe from a single point at the beginning of time
Comets	Small objects that orbit the Sun in highly elliptical orbits
Dwarf planets	Smaller objects and asteroids which orbit the Sun, e.g. Pluto
Earth	Our planet, orbits round the Sun approximately every 365 days
Galaxy	A group of billions of stars
Light-year	Distance travelled in space by light in one year
Milky Way	Our galaxy containing the Solar System
Moon	Natural satellite orbiting the Earth in approximately 1 month
Natural satellites	Moons which orbit planets
Orbital period	Time for an object in space to complete one orbit
Phases of the Moon	Changes in appearance of the Moon as it orbits the Earth
Planets	Eight large objects, which orbit the Sun
Redshift	Increase of wavelength of light from a receding star or galaxy
Solar System	The Sun, the eight major planets, minor planets, asteroids and other bodies
Stars	Consist mainly of hydrogen and helium, release energy from nuclear fusion reactions
Sun	Our star at the centre of the Solar System
Universe	Hundreds of billions of galaxies that make up all the matter that exists
CMBR	Cosmic microwave background radiation
Hubble constant (H_0)	Symbol H_0, ratio of the speed at which a galaxy is moving away from the Earth to its distance from the Earth
Red giant	State of most stars near the end of their life
Red supergiant	State of largest stars near the end of their life before they explode as supernovas

6.1 Earth and the Solar System

6.1.1 The Earth

Key objectives

By the end of this section, you should be able to:
- know how the Earth orbits the Sun
- explain how the tilt of the Earth's axis of rotation causes the seasons

- know that the Moon orbits the Earth in about one month and use this to explain the Moon's phases.

- define average orbital speed and recall and use the equation: $v = 2\pi r/T$

Motion of the Earth

The **Earth** is a planet which rotates on its axis once in approximately 24 hours. The **Sun** appears to move when observed from a point on the Earth's surface. Figure 6.1 is a simplified view, not to scale, from above the North Pole of the Earth of how the Sun is seen at a point labelled O at 6 am, noon, 6 pm and midnight.

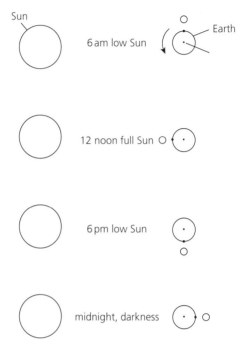

▲ **Figure 6.1 The Sun as seen from the Earth**

When point O is on the side of the Earth facing the Sun, it is day. Then point O moves to the side of the Earth away from the Sun and it is night.

The axis of the Earth's rotation is not at right angles to the plane of its rotation around the Sun but it is tilted at an angle of about 20°, as shown by Figure 6.2.

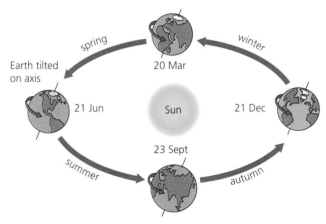

▲ **Figure 6.2 The tilt of the Earth's axis**

It can be shown that because of this tilt, the lengths of day and night are not equal and vary during the year.

The Earth orbits the Sun once in approximately 365 days. This together with the tilt of the axis of rotation explains the seasons. Figure 6.3 shows the seasons for the Southern hemisphere.

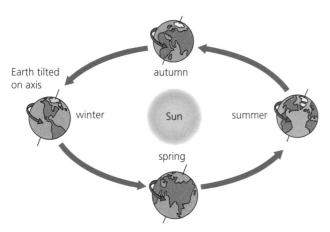

▲ **Figure 6.3 Seasons for the Southern hemisphere**

From autumn through winter to spring, the Southern hemisphere is tilted away from the Sun so it receives daylight for a shorter time every day. It receives less solar radiation so it is cooler.

Motion of the Moon

The **Moon** takes about one month to orbit the Earth. As the moon orbits the Earth, its appearance changes between full circle and thin crescent, with other shapes in between. These are called the **phases of the Moon**.

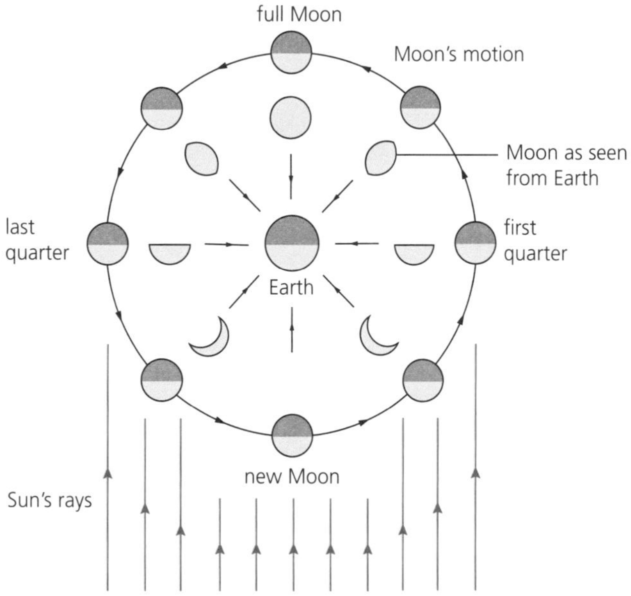

▲ **Figure 6.4 Phases of the moon**

Figure 6.4 shows how parts of the Moon's surface are illuminated by the Sun to change the appearance of the Moon.

The average orbital speed v of any object is given by the equation $v = 2\pi r/T$ where r is the average radius of the orbit and T the **orbital period**. This equation applies to all orbits, e.g. the Earth or any other planet around the Sun, the Moon around the Earth or any moon around its planet.

Revision activity

Draw the appearance of the Moon in two different phases.

Revision activity

The phases of the Moon

You should prepare yourself to observe the Moon every day for a month. Initially you will observe just after sunset and later it will have to be shortly before dawn. It will not matter if you miss a few days. If you live in a part of the Earth where the weather is cloudy, you will miss a few or possibly several observations.

● Find out the date of the next new Moon from an app, the internet or your teacher.

● One or two days after that, the Moon will be visible as a very thin crescent shortly after sunset. Sketch the appearance of the Moon and record the number of days after the new Moon you made the observation.

● Continue to observe and record the appearance of the Moon every day for a month if you can. You will have to be flexible with the time you observe. When the Moon becomes fuller than a semicircle, it is often possible to see it during daylight.

● By observing regularly, you will discover the pattern and timing of the Moon's phases. Maybe it will help to repeat for another month later using your experience of the first attempt to time your observations better.

● Compare your observations with Figure 6.4 and note any comments you have about this comparison.

Skills

Rearranging the orbital speed equation

You need to be able to rearrange the equation $v = 2\pi r/T$ to make any of the variables the subject.

● Write down the equation when r is the subject.

$$v = \frac{2\pi r}{T}$$

$$\frac{2\pi r}{T} = v$$

$$2\pi r = vT$$

$$r = \frac{vT}{2\pi}$$

● Write down the equation when T is the subject.

$$v = \frac{2\pi r}{T}$$

$$\frac{2\pi r}{T} = v$$

$$2\pi r = vT$$

$$vT = 2\pi r$$

$$T = \frac{2\pi r}{v}$$

Sample questions

REVISED

1 Draw a line from each description in the left column to one of the time durations in the right column.

Description	Time duration
hours of daylight in every 24 hours in spring	12 hours
time taken by the Earth to orbit the Sun	24 hours
time taken by the Moon to orbit the Earth	1 month
time taken by the Earth to turn one revolution on its axis	365 days

[4]

Student's answer

Description	Time duration
hours of daylight in every 24 hours in spring	12 hours
time taken by the Earth to orbit the Sun	24 hours
time taken by the Moon to orbit the Earth	1 month
time taken by the Earth to turn one revolution on its axis	365 days

[3]

Teacher's comments

The student has failed to follow instructions and drawn two lines from 'time taken by the Earth to turn one revolution on its axis' on the left and no line from 'hours of daylight in every 24 hours in spring' on the left.

Correct answer

Description	Time duration
hours of daylight in every 24 hours in spring	12 hours
time taken by the Earth to orbit the Sun	24 hours
time taken by the Moon to orbit the Earth	1 month
time taken by the Earth to turn one revolution on its axis	365 days

[4]

2   The orbital speed of the Earth around the Sun is 30 000 m/s. Calculate the average radius of the Earth's orbit.   [4]

Student's answer

$$r = \frac{v}{2\pi T}$$

$$r = \frac{3 \times 10^4}{365 \times 2\pi} = 13 \text{ m}$$

[1]

Correct answer

$$v = \frac{2\pi r}{T}$$

$$r = \frac{v}{2\pi T}$$

$T = 365 \text{ days} = 365 \times 24 \times 3600 \text{ s}$

$v = 3.0 \times 10^4 \text{ m/s.}$

$$r = \frac{3.0 \times 10^4 \times 365 \times 24 \times 3600}{2\pi} = 1.5 \times 10^{11} \text{ m}$$

[4]

Teacher's comments

The student incorrectly rearranged the equation and substituted a time in days not seconds. The student should have considered the answer and realised that 13 m was quite impossible for the radius of the Earth's orbit.

6.1.2 The Solar System

Key objectives

By the end of this section, you should be able to:
- describe the Solar System as containing the Sun, the eight named planets including their order from the Sun, minor planets and other bodies

  - understand the elliptical nature of orbits around the Sun

- know the different compositions of the inner and outer planets and explain this using the accretion model

  - analyse and interpret planetary data

- know how the strength of a gravitational field of an object depends on the mass of the object and the distance from the object
- calculate the time it takes for light to travel between objects in the Solar System
- know that the Sun contains most of the mass of the Solar System and the force that keeps objects in orbit around the Sun is gravitational attraction

  - know that orbital speeds of planets vary with distance from the Sun and position in an elliptical orbit

The Solar System

The Sun is a **star** at the centre of the **Solar System**.

The main objects of the Solar System orbiting the Sun are shown in Figure 6.5.

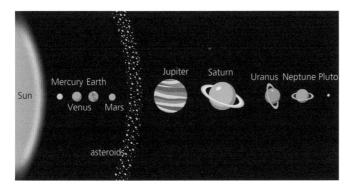

▲ **Figure 6.5 The Solar System (distances from the Sun not to scale)**

The first four **planets** in order of distance from the Sun are rocky and small: Mercury, Venus, Earth and Mars.

The next four planets in order of distance from the Sun are gaseous and large: Jupiter, Saturn, Uranus and Neptune.

There are also smaller Solar System bodies: asteroids, **comets** and **dwarf planets** such as Pluto which may orbit in the asteroid belt or further from the Sun than Neptune.

Many of the planets have moons or **natural satellites** which orbit around them.

All objects orbiting the Sun have elliptical orbits, although some are approximately circular. Except in the cases of circular orbits, the Sun is not at the centre of the orbit.

▼ Table 6.1 Data for the planets

Planet	Av distance from Sun/ million km	Orbit time around Sun/ days or years	Surface temperature/ °C	Density/ kg/m³	Diameter/ 10³ km	Mass/10²⁴ kg	Gravitational field strength/ N/kg	Number of moons
Mercury	57.9	88 d	350	5427	4.8	0.330	3.7	0
Venus	108.2	225 d	460	5243	12.1	4.87	8.9	0
Earth	149.6	365 d	20	5514	12.8	5.97	9.8	1
Mars	227.9	687 d	−23	3933	6.8	0.642	3.7	2
Jupiter	778.6	11.9 y	−120	1326	143	1898	23.1	79
Saturn	1433.5	29.5 y	−180	687	120	568	9.0	82
Uranus	2872.5	84 y	−210	1271	51	86.8	8.7	27
Neptune	4495.1	165 y	−220	1638	50	102	11.0	14

The four planets closest to the Sun, Mercury, Venus, the Earth and Mars, are rocky and smaller than the outer four planets. You can see in Table 6.1 their smaller diameter and higher density. The outer four planets are not only much larger but also have a lower density as they are gaseous.

The **accretion model** for the Solar System explains these differences.

- The Solar System was formed about 4.5 billion years ago from clouds of hydrogen gas and dust and heavier elements produced from a supernova which exploded.
- Hydrogen and some helium were drawn together by gravitational attraction to form the Sun.
- The remaining small particles joined together to form a disc in an accretion process as the material rotated.
- All the planets orbit the Sun in the same direction and lie in roughly the same plane, which is only likely if they were all in the accretion disc.
- As the Sun grew in size and temperature, light molecules such as hydrogen could not exist in a solid state.
- Heavier elements gradually accreted by gravitational attraction to grow into the inner planets.
- The lighter elements drifted further from the Sun and eventually grew by gravitational attraction to be large enough to attract even the lightest elements to form the gaseous outer planets.

Gravitational field strength of a planet

Table 6.1 shows that the gravitational field strength at the surface of a planet depends on the mass of the planet and its density.

The strength of the gravitational field of any object decreases as the distance from the object increases. So the gravitational field around a planet decreases as the distance from the planet increases. Most of the mass of the Solar System is in the Sun, so planets orbit the Sun, kept in orbit by gravitational attraction of the Sun.

Revision activity

Work in pairs. Choose a planet for your partner. Your partner chooses a different planet for you. Use Table 6.1 to look up the average distances of your two planets from the Sun. Each of you calculate the time it would take for light from the Sun to reach your planet. When you have finished, compare your answers and check for consistency.

The strength of the Sun's gravitational field decreases as the distance from the Sun increases. The orbital speeds of the planets decrease as the distance from the Sun increases. This can be confirmed by calculation using the data from Table 6.1.

The time it takes for light to travel between the objects can be calculated using the velocity equation from Topic 1 and the velocity of light 3.0×10^8 m/s from Topic 3.

When an object travels in an elliptical orbit around the Sun, its velocity is greater when it is closer to the Sun. The gravitational potential energy is greater the further the object is from the Sun. However, its kinetic energy increases, as the velocity increases when it is closer to the Sun. Energy is conserved so the total energy is the same at all points of the orbit.

Figure 6.6 shows a comet moving round the Sun in an elliptical orbit. At point X, the gravitational potential energy is least, and the velocity and kinetic energy are greatest. At point Y, the gravitational potential energy is greatest, and the velocity and kinetic energy are least.

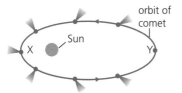

▲ **Figure 6.6 Elliptical orbit of a comet**

Revision activity

Write down the energy changes of a comet as it moves from its furthest to its closest distance from the Sun.

Sample questions

REVISED

3 What are the main common characteristics of the composition and size of the inner four planets? [2]
4 What are the main common characteristics of the composition and size of the outer four planets? [2]
5 State two objects other than the eight main planets which orbit the Sun. [2]

Student's answers

3 Rocky and hot	[1]
4 Gaseous and far from the Sun	[1]
5 Comets and satellites	[1]

Correct answers

3 Rocky and relatively small	[2]
4 Gaseous and relatively large	[2]
5 Comets and dwarf planets	[2]

Teacher's comments

3 Rocky is correct; hot is not about the composition or size.
4 Gaseous is correct; far from the Sun is not about the composition or size.
5 Students should write any **two** of: dwarf planets, comets or asteroids.

6 This question is about the accretion model for the formation of the Solar system.
   a Which materials accreted to form the Solar System? [2]
   b Which force caused this accretion? [1]
   c Explain why it is likely that the material was in a rotating disc at one stage. [2]

Student's answers

a Hydrogen and plutonium	[1]
b Electrostatic attraction	[0]
c All of it goes the same way round.	[1]

Teacher's comments

a Hydrogen is correct; plutonium incorrect.
b Incorrect answer.
c 1 mark awarded for the right idea of orbiting in the same direction.

Correct answers

    a Hydrogen, dust and heavier elements   [2]

    b Gravitation   [1]

    c All the planets orbit roughly in a plane and in the same direction.   [2]

7 a State with a reason how the gravitational fields on the surface of Mars and Neptune compare.   [2]

    b State with a reason how the orbital speeds of Mars and Neptune compare.   [2]

    c Confirm your answer to part **a** using the data from Table 6.1.   [3]

Student's answers

    a *Neptune because it is bigger.*   [0]

    b *The orbital speed of Mars is greater because its year is shorter.*   [1]

    c *orbital speed of Mars = 24 000 m/s*

       orbital speed of Neptune = 2700 m/s

       so orbital speed of Mars is greater   [1]

Teacher's comments

    a The student possibly meant that the gravitational field on the surface of Neptune was larger, but did not say so. The size of the planet is not the decisive factor.

    b Correct answer about orbital speed, but the orbit time around the Sun follows from the speed and distance and is not the reason.

    c The orbital speed of Mars is correct. The candidate possibly forgot the '2' in calculating the orbital speed of Neptune, but with no working no partial credit can be given. As it is based on a wrong number, the comment about the greater orbital speed of Mars is invalid.

Correct answers

    a The gravitational field on Neptune is greater because it more massive.   [2]

    b The orbital speed of Mars is greater because it is closer to the Sun.   [2]

    c orbital speed of Mars $= \dfrac{2\pi \times 230 \times 10^9}{687 \times 24 \times 3600} = 24000\,\text{m/s}$

       orbital speed of Neptune $= \dfrac{2\pi \times 4500 \times 10^9}{165 \times 365 \times 24 \times 3600} = 5400\,\text{m/s}$

       So the orbital speed of Mars is greater.   [3]

8 a A student states that it takes light 8 minutes to travel from the Sun to the Earth. Using data from Table 6.1, calculate the velocity of light based on this time. Note: Extension students need to know the value of the velocity of light in a vacuum. [3]
   b Comment on the accuracy of this value of the velocity of light and the student's statement. [1]

   c The average distance from the Earth to Neptune is 4500 million km and the velocity of light is 3×10^8 m/s. Calculate the time in hours light takes to travel from Neptune to the Earth. [2]

Student's answers

a velocity = distance/time = $1.5 \times 10^{11}/8 \times 60 = 3.1 \times 10^5$ m/s [2]
b They are nearly right. [0]

c time = distance/velocity = $4.3 \times 10^{12}/3.0 \times 10^8 = 1.4 \times 10^4$ s = 4.0 h [1]

Teacher's comments

a The student made an error of factors of 10 in the calculation, so loses a mark.
b As the comment is based on a wrong calculation, no mark is scored.

c The student used a slightly wrong value in the correct equation.

Correct answers

a velocity = distance/time = $1.5 \times 10^{11}/8 \times 60 = 3.1 \times 10^8$ m/s [3]
b The velocity of light is close to the correct value, so the time for light to travel from the Sun is fairly accurate. [1]

c time = distance/velocity = $4.5 \times 10^{12}/3.0 \times 10^8$
              $= 1.5 \times 10^4$ s = 4.2 h [2]

9 Figure 6.7 shows the elliptical orbit of an asteroid.

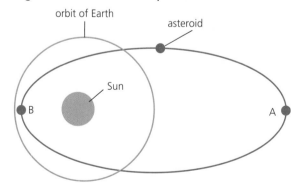

▲ **Figure 6.7**

Describe the change of the following quantities as the asteroid moves from point A to point B:
distance from the sun, velocity, gravitational potential energy, kinetic energy, total energy [5]

Student's answer

distance from the sun decreases, velocity increases, gravitational potential energy increases, kinetic energy increases [3]

Teacher's comments

Correct answers for distance from the Sun, velocity and kinetic energy. The answer for gravitational potential energy is incorrect and the student failed to write anything about total energy.

Correct answer

distance from the sun decreases, velocity increases, gravitational potential energy decreases, kinetic energy increases, total energy stays the same [5]

Revision activity

Make flash cards to revise and rearrange the equation connecting orbital speed, radius of orbit and orbital period. Include what the symbols in the equation represent and rearrange the equation with r on its own on the left of the equals sign and then with T on its own on the left.

Revision activity

Observing planets in the night sky

Three of the planets Venus, Jupiter and Mars appear as bright 'stars' and can easily be observed at the right time of year and time of day. Saturn and Mercury can also be seen with the naked eye but can be hard to see, especially Mercury.

You should prepare yourself to observe the planets every week for a few months. Again, if you live in a part of the Earth where the weather is cloudy, you will miss a few or possibly several observations.

● Find out which of Venus, Jupiter and Mars are visible where you live and at what time of night from an app, the internet or your teacher.
● See if you can observe these planets at times that are reasonable for you. Don't let it disturb your sleep!
● Make notes and record the date and time of observing any planet. Try to record where you observed it in the sky and how high up.

Exam-style questions

Answers available at: www.hoddereducation.co.uk/cambridgeextras

1 a Sketch a labelled diagram to show:
      i   the orbit of the Earth around the Sun   [2]
      ii the Earth's axis of rotation   [1]
   b Describe the angle of the axis of rotation.   [1]

2 In the table below, which describes the planet Venus?

A	gaseous and large
B	gaseous and small
C	rocky and large
D	rocky and small

                                                                                   [1]

3 In the table below, which compares properties of the Earth and Jupiter?

	Property of the Earth	Property of Jupiter
A	relatively low density	relatively high density
B	relatively low mass	relatively high mass
C	relatively high diameter	relatively low diameter
D	relatively high density	relatively low mass

                                                                                   [1]

4 a The planet Mercury has an orbital radius of 58 million km and an orbital speed of 48 000 m/s.
      Show that the orbital period of Mercury is 88 days.   [3]
   b Without calculation, write down how the following properties for the planet Uranus compare with the properties for the planet Mercury:
      i   orbital speed   [1]
      ii strength of the Sun's gravitational field   [1]

5 Using data from Table 6.1:
   a State and explain the differences of orbit time, average distance from the Sun and surface temperature of the planets Uranus and Neptune.   [4]
   b Considering density, diameter and mass, explain the difference between gravitational field strength on the surface of the planets Mars and Jupiter.   [2]

6 a State the differences in nature of the four inner planets in the Solar System from the four outer planets and why there are these differences.   [2]
   b Use the accretion model to explain why there are these differences.   [4]

7 Light travels at 3.0×10^8 m/s in space. The distance of the Moon from the Earth is 390 000 km. Calculate the time it takes for light to travel from the Moon to the Earth.   [3]

8 a Make a labelled sketch of the typical orbit of a comet going round the Sun. Label two points X and Y at two points on the orbit.   [3]
   b State the difference in velocity, gravitational potential energy and total energy between X and Y.   [4]

6.2 Stars and the Universe

6.2.1 The Sun as a star

Key objectives

By the end of this section, you should be able to:
- know that the Sun is a medium-sized star containing mostly hydrogen and helium
- the Sun radiates energy mostly in the infrared, visible and ultraviolet regions of the electromagnetic spectrum

- know that stars are powered by nuclear fusion reactions with hydrogen

The Sun is a medium-sized star. It is made up almost entirely of hydrogen: there is some helium and other elements. It radiates energy over the whole electromagnetic spectrum but mostly in the infrared, visible and ultraviolet regions.

Stars are powered by nuclear fusion reactions. Stable stars produce helium from the fusion of hydrogen.

Sample question

10 a What two elements are the main constituents of the Sun? [2]

  b Name one of the main types of radiation from the Sun in addition to visible light. [1]

  c Describe the nuclear reaction that takes place in the Sun. [3]

Student's answers

  a hydrogen and deuterium [1]

  b X-rays [0]

  c Nuclear explosion of hydrogen. [1]

Teacher's comments

  a Hydrogen is correct, but deuterium is an isotope of hydrogen. A different element was required.

  b X-rays are given off by the Sun, but the main electromagnetic radiations in addition to visible light are infrared and ultraviolet.

  c This answer is much too vague.

Correct answers

  a hydrogen and helium [2]

  b infrared (or ultraviolet) [1]

  c nuclear fusion of hydrogen to form helium [3]

6.2.2 Stars

Key objectives

By the end of this section, you should be able to:
- know that galaxies are groups of hundreds of billions of stars and the Solar System is in a galaxy called the Milky Way
- be able to calculate the value in metres of the astronomical distance 1 light-year, the distance light travels in a year

- know that 1 light-year $= 9.5 \times 10^{15}$ m
- describe the life cycle of a star

Galaxies

Galaxies are groups of hundreds of billions of stars. The Solar System is in a galaxy called the **Milky Way**. The other stars of the Milky Way are much further from the Solar System than the distances between the Sun and the planets of the Solar System.

An astronomical unit of distance is the **light-year**, the distance travelled in space by light in one year.

One light year is equal to 9.5×10^{15} m.

Skills

Calculating the value of the astronomical distance 1 light-year in metres

$$distance = speed \times time$$
$$= 3 \times 10^{8} \times 365 \times 24 \times 3600$$
$$= 9.5 \times 10^{15} \text{ m}$$

Revision activity

Make flash cards to revise information about galaxies. Include the name of the galaxy containing the Earth, how many stars are in this galaxy, the diameter of this galaxy and how many galaxies are in the Universe.

Life cycle of stars

The life cycle of a star shown in Figure 6.8.

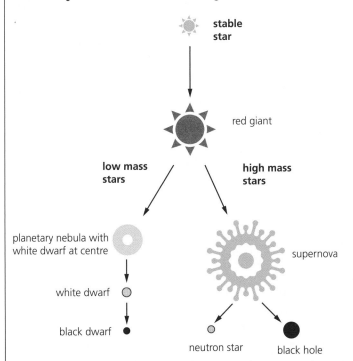

▲ Figure 6.8 Life cycle of stars

- The materials which form stars are interstellar clouds of gas and dust containing hydrogen.
- The interstellar clouds collapse due to gravitational attraction, increase in temperature and form protostars.

- Stable stars form from protostars when the inward gravitational forces balance the outward forces as the temperature of the star increases. The high kinetic energy at high temperatures causes a large pressure, producing the outward forces.
- All stars eventually use up all the hydrogen for fusion reactions.
- When this happens, most stars become **red giants**.
- Red giants become planetary nebulas with a white dwarf at the centre.
- Massive stars expand to form **red supergiants**.
- A red supergiant explodes as a supernova, forming a nebula of hydrogen and new heavier elements created in the explosion.
- The explosion of a supergiant leaves behind a neutron star or a black hole.
- The nebula from a supernova may form new stars with orbiting planets.

Sample questions

11 Calculate the distance of a light-year in metres. The velocity of electromagnetic waves in a vacuum is 3.0×10^8 m/s. [4]

Student's answer

light-year $= 3.0 \times 10^8 \times 365 \times 3600 = 3.9 \times 10^{14}$ m [3]

Note: Extension students are expected to know that 1 light-year $= 9.5 \times 10^{15}$ m and that the velocity of electromagnetic waves in a vacuum is 3.0×10^8 m/s.

Correct answer

One light-year = distance travelled by light in a year

$$= v \times t$$

$$= 3.0 \times 10^8 \times 365 \times 24 \times 3600 = 9.5 \times 10^{15} \text{ m}$$ [4]

Teacher's comments

The student used the correct equation correctly, but left out the factor 24 for the number of hours in a day.

12 Choose words from the following list that describe what stars of normal size can become or produce after using up all their hydrogen.
red giant   red supergiant   white dwarf   red dwarf
heavier elements   neutron star   black hole [2]
13 Choose words from the following list that describe what massive stars much greater than normal size can become after using up all their hydrogen.
red giant   red supergiant   white dwarf   red dwarf
heavier elements   neutron star   black hole [3]

Student's answers

12 <u>red giant</u>   red supergiant   <u>white dwarf</u>   red dwarf
   <u>heavier elements</u>   neutron star   black hole   [1]

13 red giant   <u>red supergiant</u>   white dwarf   red dwarf
   <u>heavier elements</u>   <u>neutron star</u>   <u>black hole</u>   [2]

Correct answers

12 <u>red giant</u>   red supergiant   <u>white dwarf</u>   red dwarf
   heavier elements   neutron star   black hole   [2]

13 red giant   <u>red supergiant</u>   white dwarf   red dwarf
   <u>heavier elements</u>   <u>neutron star</u> OR <u>black hole</u>   [3]

Teacher's comments

12 Red giant and white dwarf are correct answers, but the only correct answers. Heavier elements are not formed in stars of normal size. One mark is lost for the incorrect additional answer.

13 Red supergiant and heavier elements are correct answers. Either a neutron star or a black hole is formed. One mark is lost for stating both.

6.2.3 The Universe

Key objectives

By the end of this section, you should be able to:
- know that the Milky Way is one of hundreds of billions of galaxies in the Universe and the diameter of the Milky Way is about 100 000 light-years
- describe redshift as an increase of wavelength of light from receding stars and galaxies compared with light from a stationary source
- know that redshift is evidence for the Big Bang Theory
- explain what is meant by cosmic microwave background radiation (CMBR)
- know how the speed at which a galaxy is moving away from Earth and the distance to a far galaxy can be found
- define the Hubble constant H_0 and know its current estimated value is $2.2 \times 10^{-18}\,s^{-1}$
- know the equation containing the Hubble constant for the age of the Universe.

The Universe

The Milky Way is one of hundreds of billions of galaxies in the **Universe**. The Milky Way is one large disc with spiral arms of diameter about 100 000 light-years, containing hundreds of billions of stars. The Solar System is in one of the minor spiral arms.

Figure 6.9 shows a simplified diagram of the Milky Way galaxy.

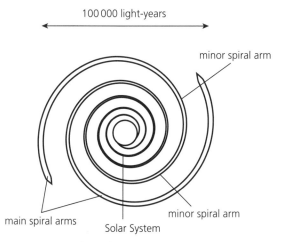

▲ **Figure 6.9 The Milky Way**

The expanding Universe and the Big Bang Theory

Nearly all the stars and galaxies of the Universe are moving away from the Earth at high speed. This increases the wavelength of light and other electromagnetic radiation observed on Earth emitted from receding stars and galaxies, so the light appears redder than when emitted. This is called **redshift** and is illustrated in Figure 6.10.

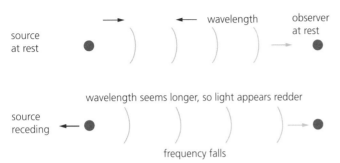

▲ **Figure 6.10 Redshift**

Redshift of this light shows that the Universe is expanding. This is consistent with the **Big Bang Theory** that the Universe was initially packed in a single point, which exploded over 10 billion years ago.

> Electromagnetic radiation from the Big Bang is still observed in the microwave region today, everywhere in space. It was produced shortly after the Universe was formed and this radiation has been expanded into the microwave region of the electromagnetic spectrum as the Universe expanded. It is called cosmic microwave background radiation (**CMBR**).
>
> The speed (v) of a galaxy's recession from the Earth can be worked out by measuring the amount of redshift. The brightness of a supernova in a far galaxy enables its distance (d) from the Earth to be worked out.
>
> The ratio of v and d equals the **Hubble constant (H_0)** as given by the equation:
>
> $$H_0 = \frac{v}{d}$$
>
> The value of H_0 is currently estimated to be $2.2 \times 10^{-18}\,\text{s}^{-1}$.

Skills

Rearranging the equation with the Hubble constant
You need to be able to rearrange the equation
$H_0 = v/d$ to make any of the variables the subject.
● Write down the equation when v is the subject:

$$H_0 = \frac{v}{d}$$

$$\frac{v}{d} = H_0$$

$$v = dH_0$$

● Write down the equation when d is the subject:

$$H_0 = \frac{v}{d}$$

$$dH_0 = v$$

$$d = \frac{v}{H_0}$$

Age of the Universe

An estimate for the age of the Universe from when all matter was at a single point is given by d/v. So rearranging the previous equation gives:

$$\frac{d}{v} = \frac{1}{H_0} = \text{age of the Universe}$$

Skills

Rearranging the equation for the age of the Universe

You need to be able to rearrange the equation for the age of the Universe in terms of v and d to make any of the variables the subject.

$$\frac{d}{v} = \text{age of Universe}$$

● Write down the equation when d is the subject:

$$d = v \times \text{age of Universe}$$

● Write down the equation when v is the subject:

$$d = v \times \text{age of Universe}$$

$$v \times \text{age of Universe} = d$$

$$v = \frac{d}{\text{age of Universe}}$$

Sample questions

REVISED

14 A galaxy is 40 million light-years distant from the Solar System. Calculate the speed it is moving away from the Earth. [3]

Student's answer

$v = H_0 d = 2.1 \times 10^4 \, \text{m/s}$ [1]

Correct answer

$v = H_0 d = 2.2 \times 10^{-18} \times 40 \times 10^6 \times 9.5 \times 10^{15} = 8.4 \times 10^5 \, \text{m/s}$ [3]

15 Use the value of the Hubble constant to estimate the age of the Universe in billions of years. [3]

Student's answer

$\text{age of the Universe} = 1/H_0 = 1/2.2 \times 10^{-18} = 4.5 \times 10^{17} \, \text{s}$ [1]

Correct answer

$$\text{age of the Universe} = 1/H_0 = 1/2.2 \times 10^{-18} = 4.5 \times 10^{17} \, \text{s}$$

$$= \frac{4.5 \times 10^{17}}{365 \times 24 \times 3600}$$

$$= 1.4 \times 10^{10} \text{ years}$$

$$= 14 \text{ billion years} \quad [3]$$

Teacher's comments

The answer is too small by a factor of 4. Possibly the candidate forgot the number 40 and also made a factor of 10 error. The examiner cannot see what the candidate did, so is not able to give any credit beyond that for the correct equation.

Although the basic mathematics is straightforward, the working out is often complex in Space physics questions. Candidates for the Extended paper are expected to be able to do such calculations.

Teacher's comments

The student's answer is correct as far as it goes, but is incomplete. The question asked for the answer in billions of years not in seconds.

Revision activity

Make flash cards to revise about CMBR. Include what CMBR stands for and what CMBR means in relationship to the Big Bang.

Exam-style questions

Answers available at: www.hoddereducation.co.uk/cambridgeextras

9 In the table below, which is the name of the Sun's galaxy and the distance of most other stars in the galaxy from the Sun compared with the distance of the Earth from the Sun?   [1]

	Name of the Sun's galaxy	Distance of other stars in the galaxy from the Sun
A	Solar System	about the same
B	Milky Way	much further
C	Solar System	much closer
D	Milky Way	about the same

10 a State how the size of the Sun compares to most stars in the Universe.   [1]
   b State which elements make up most of the Sun.   [2]
   c State what powers the Sun.   [2]
   d State what is contained in a galaxy.   [1]
   e State the name of the galaxy where the Earth is situated.   [1]
11 The diameter of the Milky way is 100 000 light years. Calculate the diameter of the Milky Way in metres.   [4]
12 a State what is redshift.   [2]
   b Explain the connection between redshift and the Big Bang Theory. [2]

13 Choose from the table below what most stars form as the next stage in their life cycle after running out of hydrogen.

A	red giant
B	red supergiant
C	white dwarf
D	neutron star

[1]

14 a What do the letters CMBR stand for.   [2]
   b State and explain if CMBR radiation is part of the electromagnetic spectrum.   [2]
   c Explain the connection between CMBR and the Big Bang Theory.   [2]
   d Write down the distance of 1 light-year in metres.   [1]
15 a Explain how the speed of a galaxy moving away from the Earth can be determined.   [2]
   b State how the distance away from the Earth of a far galaxy can be determined.   [2]
16 a Write down an equation which defines the Hubble constant. Use the normal symbols for all quantities in your equation and explain what these quantities represent.   [5]
   b Write down the current estimate for the value of the Hubble constant and its unit.   [2]
17 a Write down the equation which estimates the age of the Universe. Use the normal symbols for all quantities in your equation and explain what these quantities represent.   [5]
   b Work out an estimate for the age of the Universe in years.   [3]
   c Write down a statement about all the matter in the Universe being at a single point.   [2]

Index

Note: page numbers in **bold** refer to the location where a key definition is *first* defined.

© *Mike Folland and Catherine Jones 2022*